Practical Soft Systems Analysis

Practical Soft
Systems Analysis

David Patching
FMS EuroIE

PITMAN
PUBLISHING

PITMAN PUBLISHING
128 Long Acre, London WC2E 9AN

A Division of Longman Group UK Limited

© D Patching 1990

First published in Great Britain 1990
Reprinted 1993
British Library Cataloguing in Publication Data

Patching, David
  Practical soft systems analysis.
  1. Data processing. Systems analysis
  I. Title
  004.21

  ISBN 0-273-03237-2

Printed in England by Clays Ltd, St Ives plc

Contents

To Will and Annie Dickson

Foreword

Systems thinking is a powerful approach to dealing with modern day complexities. The basic idea is that we consider how things might operate together as a whole, as well as accepting the value of traditional reductionist thought. This breakthrough might seem fundamental, but it was not until the 1930s or later that science woke up to the idea in biological studies, and the 1940s before it was applied with any significance in technology and management. The systems idea has, however, been around in philosophy for many centuries, from ancient Greeks to more recent occidental thinkers such as Immanuel Kant and Karl Marx, and similarly onwards from ancient oriental philosophers.

Nowadays it is hard to escape the effect of systems thinking. Just about every discipline has felt the influence, pervading University-based activities and those of practitioners from a variety of professions. The ideas have been used for *scientific* heuristic studies and the evolution of knowledge, and have been shown to have practical value for planners, decision makers, managers and problem-solvers generally. There are many approaches that use the systems perspective for addressing problems; for example systems engineering, system dynamics, viable system diagnosis, strategic assumption surfacing and testing, interactive planning, critical systems heuristics, and the soft systems methodology. Each has strengths in either design, prediction, strategic organisation, idealised planning, social planning or strategic options analysis, bearing in mind that these strengths will only hold true for the particular type of difficulty being examined.

In this book, David Patching has prepared an account of systems-based problem-solving specifically from a *practitioner's* point of view, choosing to look in detail at the *soft systems methodology,* as developed by staff and students at the University of Lancaster in the United Kingdom. The leading protagonist, Peter Checkland, hit the systems community with one main idea, ie that social dynamics are better understood in terms of the *subjective* nature of human interpretation, whereas in the past systems thinking failed to reason beyond a primitive and *impersonal* understanding of natural and social functional units. Checkland and colleagues reinterpreted the systems idea through a programme of action research, and the methodology that evolved from these efforts brings to the fore the viewpoints of those who are immersed in an organisation's difficulties. Rigorously working through considerations of these viewpoints brings a richness to the analysis that could never be attained by a group of external 'experts'. The number of inapt solutions that I have witnessed when working in all sorts of practical situations is a testament to this; in my opinion, those involved in a problem situation are far better qualified in many respects for thinking of ways out of the difficulties that they are facing. The real value of the approach lies in the way the variety of *viewpoints* are accepted as a strength and are positively harnessed, and as a device for *learning* about a situation where the definition of strategic options is itself problematic.

The soft systems methodology, therefore, is a novel and extremely useful way of carrying out an analysis of strategic options, as it aims to promote meaningful debate among those actually facing the difficulties; debate about structural, procedural and attitudinal changes. However, because of the relative newness of systems ideas, it is sometimes difficult to persuade pragmatic managers of the value of adopting and making use of what could be considered a wholly different approach to both *management* and *problem-solving*. It is to the credit of David Patching that he has recognised the potential of soft systems thinking and introduced the methodology into his own organisational context. Not content with this achievement, he has also made a genuine attempt to spread this kind of thinking throughout his working discipline, including the development and design of seminars held on behalf of the Institute of Management Services, with which I have had an invited role. The writing of the volume that follows is yet another substantial effort to disseminate the ideas to a much broader audience of practising people.

David's main concern in his work is to develop a pragmatic account of soft systems ideas, rather than being pedantic about academic issues. In this book, he has put across the ideas in an accessible and usable style, directed at opening up these ideas to colleagues and the like who address similar difficulties at the sharp end of problem-solving practice. In this commendable endeavour, I am pleased to have been offered a role.

Robert L Flood
University of Hull
United Kingdom
June 1990

Preface

In the words of Bob Dylan, *"the times they are a' changing"*, and changes of any sort are seldom isolated in their effects. In the last decade new technology has led to significant improvements in communication methods, shrinking the world to size and making it possible for everyone to know what is happening at any moment from Darwin to Derby, or from Moscow to Manchester. At the same time, and partly as a result, fundamental changes have been taking place to the political make-up of countries and continents, economic strategies, and the cultural and sociological aspects of most nations. None of these happen in isolation from the others; new technology has improved communications, enabling political, sociological and economic ideals to be readily disseminated on a global basis, leading to changes in governments and governmental policies, which in turn affect local practices, and thereby individuals who carry them out. In the United Kingdom alone there has been a shift in emphasis towards a free-market economy, affecting both the private and public sectors, leading to a degree of interdependence where the failure of one party could have crucial effects on the other. On a global, national and even personal level, new relationships and interdependencies are rapidly being established, ones which need to be recognised before adequate solutions can be developed to the problems that will inevitably arise.

In many respects, these relationships are primarily the concern of the policy-makers. However, as the effects filter through to those who work in the problem-solving professions, the existing disciplines of systems analysis, work study, and organisations and methods will need to be supplemented by an approach that extends their horizons, and encourages a wider view to be taken than in the past. Problem solving is, of course, by no means confined to these groups, and it is equally important that managers, supervisors or anyone concerned with making improvements is able to do so in full knowledge of all the influencing factors. *Soft systems thinking* crosses traditional boundaries, not only of the separate analytical professions, but also of the organisations being examined. In this way, ideas can be developed that take account of the wider implications of the whole situation, at the same time considering the feelings and perspectives of the people involved.

Under the umbrella title of **Practical Soft Systems Analysis**, the book explains the *soft systems methodology* and explores its use in practice, discussing difficulties that can arise in the process. It also covers less formal applications of soft systems ideas, recognising that customised interpretations are sometimes necessary to take account of local factors, and the training, background and personality of the individuals involved.

In the process of applying these ideas, I have made extensive use of the two main textbooks that are currently available on the subject, ie *Systems Thinking, Systems Practice* by Peter Checkland (John Wiley and Sons Ltd, 1981), and *Systems Concepts, Methodologies and Applications* by Brian Wilson (John Wiley and Sons, 1984). Where specific passages have been quoted from these and other publications, they are acknowledged in the customary manner. Overall the explanations in the early parts of this book reflect the content of these texts, but have been given an interpretation that has proved useful in my own situation. *Dealing with Complexity* by Robert Flood and Ewart Carson (Plenum Press, 1988) has also proved invaluable, and I gratefully acknowledge the help that Bob Flood himself has given me in recent years, not only in improving my knowledge, but by encouraging my efforts to apply, teach and write about systems.

My experience in this field stems from an involvement with a European-community *ESPRIT* project, which was concerned with the development of a methodology for the *Functional Analysis of Office Requirements* (FAOR), an approach that makes extensive use of soft systems concepts. Since that time I have used these ideas, blended with systems knowledge gained from a long involvement with air engineering in the Fleet Air Arm, to address organisational and other problems. When attempting to explain how *soft systems analysis* can be used in practice, it is important to be able to understand the thinking behind the actions that were taken, consequently the practical applications discussed in the latter part of the book are all based on studies of which I have some personal knowledge. In the descriptions given, specific references to the actual organisations have been avoided, as, in some cases, this could be considered insensitive. However, each study has been well-documented should any reader require more information. Although these studies are mainly concerned with organisations in the public sector, the applications provide general examples of how the approach can be used, and should be of benefit to analysts in any sphere of activity. In particular, they should be of interest to members and associates of the Institute of Management Services (the IMS), an Institute that, in the continuing search for ways of improving the effectiveness of organisations, has wholeheartedly embraced the soft systems approach in recent years.

Finally, a few special thanks are in order; to Ron Ollive and Ron Butler for general support and down-to-earth advice on management services matters; to Alwyn Jones of City University London, and Peter Mcloughlin of the New College Durham for enhancing my knowledge of systems matters; to Jim Sellen of the IMS for his encouragement and for insisting that I contact a publisher and make a firm commitment to the book; and to all the people who have been associated with the applications of soft systems ideas that are explored in the following pages.

David Patching
Hatfield Peverel

1 Introduction

1.1 The Style of the Book

In recent years there has been a growing awareness of the work of Professor Peter Checkland and associates from the University of Lancaster in the development of the **Soft Systems Methodology** for addressing organisational issues, and the publications *Systems Thinking, Systems Practice* and *Systems Concepts, Methodologies and Applications*, referred to in the preface, summarise the comprehensive research carried out in this field. Generally the role of these publications is to explain systems concepts and how they relate to human activity, with examples of practical applications given mainly for illustrative purposes. With some exception, they do not provide detailed advice on *how* to apply systems thinking to real-life situations, or how to overcome any difficulties that arise.

There is undoubtedly a tremendous amount of interest in soft systems ideas at present, but a paucity of advice on how to apply them in practice, the benefits that can be gained by doing so, and how they complement more familiar methods of analysis. As a result, there appears to be a gap between the pragmatists who might wish to use them for everyday study or consultancy work, but have not had the benefit of a formal systems education, and those who have been involved at a more academic level in their development and application.

It is the purpose of the book to provide a bridge for what I have called the *pragmademic gap*, and remove some of the mystique that still surrounds the subject, enabling more practitioners from the problem-solving professions to understand and apply the approach. The book, therefore, addresses it from a practical viewpoint, first translating the language of systems into common parlance, then exploring each stage of the Soft Systems Methodology whilst examining problems that could occur in practice. Other applications of systems concepts are then explained, and the final section of the book provides comprehensive examples of where soft systems analysis has been used to good effect. By adopting this style, it is hoped that the book will both complement the existing texts about the subject, and contrast with them by exploring the subject from the point of view of a practitioner.

1.1.1 General Terminology

The inclusion of the term *soft systems analysis* in the title is one used after considerable thought, as it can imply that problems are broken down to lower order elements as part of the examination, whereas soft systems thinking is primarily directed at understanding the whole situation. However, it is a term that is commonly used, and as such serves to distinguish the approach from computer or *hard* systems analysis. On the assumption that there will, of necessity, be a number of phrases and words in the book that some readers will find unfamiliar, it seems preferable to use a title that will avoid any further confusion. Furthermore, although a large part of the book is devoted to explaining the theory and practice of the Soft Systems Methodology, applying systems thinking to human situations is by no means unique to this approach. Readers may be familiar with the work of Katz and Kahn (1966) who addressed the complexities of organisation as open systems, the concepts of socio-technical systems in organisations as developed by the Tavistock Institute of Human Relations, and the application of systems theory to organisations discussed by A.K.Rice in the book *Productivity and Social Organisation* (1958). Although these theories are not explored in any depth in this book, broad-based or unstructured applications of systems ideas to situations are considered in Chapter 9 (Practical Applications), warranting the use of the generic term *soft systems analysis* in the title.

The terms *analyst* and *organisation* also occur frequently throughout the book, and these also have been chosen after careful deliberation of alternatives. **Analyst** is taken to mean *all those who observe or examine a situation with a view to improving it*, thus applying not only to those employed in the analytical professions (eg consultants, management services officers, systems analysts and so on) but also managers across the full spectrum of employment. The term analyst itself, being a fairly general and neutral annotation, also makes it unnecessary to add the usual defensive remarks that preclude being accused of sexism (eg please regard all references to he/him as she/her, etc). **Organisation**, frequently used to refer to large businesses and corporations, is used here in a more universal sense to apply to any "*body of persons organised for some purpose*" (Collins Standard Dictionary), thus covering groups of any size or scope.

The value of representing situations graphically is covered at length in Chapter 5; in further support of the concept that *a picture speaks a thousand words*, I have endeavoured to make regular use of illustrations to reinforce ideas where appropriate, in the hope that it will also make the book more stimulating than it otherwise might have been.

1.1.2 The Structure of the Book

The book is essentially designed to progress gradually from the theory to the practice of systems, ie starting with an explanation of systems concepts and their value in problem-solving terms, moving through a description of the SSM, to examples of where the approach has been used to address real-life problems. The final chapters cover particular application areas, based on studies carried out in recent years. Fig 1.1 shows an overview of the book that reflects this progression, and a brief description of the contents of Chapter 2 onwards is given below.

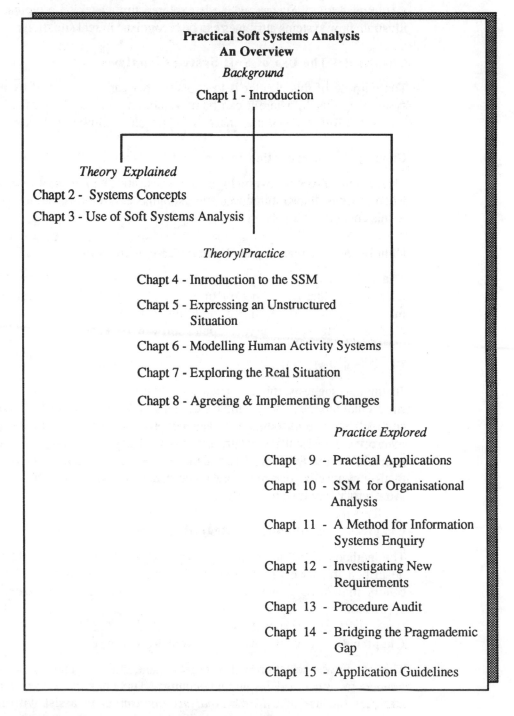

**Practical Soft Systems Analysis
An Overview**
Background
Chapt 1 - Introduction

Theory Explained
Chapt 2 - Systems Concepts
Chapt 3 - Use of Soft Systems Analysis

Theory/Practice

Chapt 4 - Introduction to the SSM

Chapt 5 - Expressing an Unstructured
             Situation

Chapt 6 - Modelling Human Activity Systems

Chapt 7 - Exploring the Real Situation

Chapt 8 - Agreeing & Implementing Changes

Practice Explored

Chapt   9 - Practical Applications

Chapt 10 - SSM for Organisational
                Analysis

Chapt 11 - A Method for Information
                Systems Enquiry

Chapt 12 - Investigating New
                Requirements

Chapt 13 - Procedure Audit

Chapt 14 - Bridging the Pragmademic
                Gap

Chapt 15 - Application Guidelines

Fig 1.1 - Overview of Practical Soft Systems Analysis

Chapter 2 - Systems Concepts

This chapter provides a summary of system types, concepts and characteristics. Examples are given of man-made systems as the basis for explaining the concept of a *Human Activity System*, which is explored further taking a typical enterprise to illustrate how systems attributes can be recognised in real situations.

Chapter 3 - The Use of Soft Systems Analysis

The purpose of this chapter is to explain what can be achieved by using the soft systems approach, where it can be of value, and, in general terms, how it differs from other forms of systems analysis and problem-solving methods.

Chapter 4 - Introduction to the Soft Systems Methodology

This chapter gives an overview of the SSM, and the *constitutive rules* that apply when using it. It also introduces and explains certain terms that may be unfamiliar in this context to the reader.

Chapter 5 - Expressing an Unstructured Situation

The initial stages of a soft systems review are covered, and the purpose, development and use of rich pictures are explored. The identification of issues in human situations is examined, and advice given on information collection and analysis, using typical microcomputer software packages.

Chapter 6 - Modelling Human Activity Systems

As the title suggests, this chapter will explain the development and construction of conceptual models, giving simple examples as the basis for understanding the more complex models introduced in later chapters. These stages of the approach are where most difficulties occur, and where many studies fail to make progress, generally due to the lack of confidence of the practitioners. The type of problems that could arise are drawn out and examined, and advice given on how to arrive at a satisfactory conclusion.

Chapter 7 - Exploring the Real Situation

The methods of making comparison between the models and the real situation are described in this chapter, covering the methods suggested by Checkland, and others that have proved useful in practice, including an *extended analysis* approach.

Chapter 8 - Agreeing and Implementing Changes

This covers the selection of feasible/desirable changes, and the courses of action needed to agree and implement change. The chapter, amongst other things, explores the use of a *Needs Analysis Instrument* to assist with assessing the attitudes of individuals and groups in relation to change. It also considers the role of internal consultants, the extent of the client's involvement in soft systems reviews, expectations in terms of the client's knowledge of the approach, and the dangers of too much *ivory-tower* thinking.

Chapter 9 - Practical Applications

The type of situations that can benefit from the use of a soft systems approach are summarised, indicating first where the full SSM could be appropriate, giving examples, and secondly, where even a limited amount of systems thinking can be of value, eg for planning purposes, in support of *hard* systems analysis, for work measurement and analysis, and the development of organisational groupings to meet new requirements. Some examples are also given of how the Soft Systems Methodology has been incorporated into computer investigation packages.

Chapter 10 - SSM for Organisational Analysis

Chapter 10 gives a comprehensive illustration of using the SSM for organisational analysis, demonstrating also its value when advising on the introduction of new technology. The study, carried out as part of a European Economic Community ESPRIT project, was undertaken in the Social Services Department of a large local authority, and highlighted a number of organisational weaklinks which had considerable influence on the subsequent investigation of new technology applications.

Chapter 11 - A Method for Information Systems Enquiry

The use of systems ideas to enquire into the information needs of organisations is explored in depth, and an example is given of how an information model can be developed. This is then used as a basis for examining one or more aspects of the organisation with a view to making improvements to the content or form of the information.The method has been specifically developed to make use of unsophisticated database software on a microcomputer to assist with information analysis.

Chapter 12 - Investigating New Requirements

The use of systems models to develop ideas about revised organisation structures to meet the requirements of new legislation is examined, exploring further the links between system structures and functional groupings in organisations, and how ideas about management information can be developed.

Chapter 13 - Procedure Audit

An example of the use of soft systems ideas to set up and implement an audit of procedures, whether they are low-level processes or activities, or high level procedures for achieving the purpose of the organisation as a whole.

Chapter 14 - Bridging the Pragmademic Gap

This chapter examines in some detail a study that combined the soft systems approach with conventional techniques for developing a user requirement for a computer system, and considers the complementary nature of soft and hard analysis, together with some pitfalls to be avoided.

Chapter 15 - Application Guidelines

In the final chapter, I have gathered together many of the points about the application of soft systems ideas that, from my personal experience, are worth bearing in mind when embarking on any project that uses this approach. Overall, I hope that readers from all areas of systems research, problem-solving and management will gain some benefit from the advice that is given in here and in the book as a whole.

2 Systems Concepts

2.1 Introduction

We are all aware of systems, either because we work with them on a day-to-day basis, or simply because the term is used frequently in everyday conversation, without consciously considering what it means in any detail. The subject of *systems* is a vast one, encompassing a wide variety of interrelated concepts, influenced in turn by the differing perspectives of system theorists, with the term itself covering a range of man-made devices, configurations that occur naturally in the world around us, and formal or informal arrangements of human beings, eg social, political, economic *systems*, and so on. My own involvement has been at two levels; initially when attempting to diagnose and repair the multitude of engineered systems that make up fixed-wing and helicopter aircraft, and, in recent years, making use of systems ideas to explore, understand and improve organisations. After this involvement and a great deal of research into the wealth of material that has been written about the subject, I have reached the conclusion that it is far from easy to provide a clear, unambiguous summary of these ideas that would also reflect the intellectual effort of the many researchers in this field.

Consequently, although this chapter aims to impart sufficient knowledge of systems concepts to put the remaining discussion into context, it is not an exhaustive coverage of these concepts; nor is it my deliberate intention to add to the systems debate or to emulate the significant achievements of the systems theorists. The content should, however, enable practitioners who have not covered systems theory in any depth to grasp the basic principles, and those who wish to explore the subject further are invited to delve into the textbooks listed in the bibliography, some of which have already been mentioned in the opening chapter. A particularly useful guide to systems ideas and how they relate to management theory is included in the Open University series of textbooks, such as *Systems, Organisations and Management* by John Beishon and colleagues (Open University Press, 1974).

Despite the variety of perspectives explored in these and other texts, there is a general consensus of agreement about the principles that underpin the soft systems approach, which are explained briefly in this chapter. It is essentially an interpretation that has a practical bias, based on a variety of projects that have used the approach to good effect in situations where problems existed. In the explanations, an example from the air engineering world is used, for which I make no apology as it provides a suitable illustration of system characteristics which are then considered in relation to human activity. The example may also give readers who dislike travelling by air something else to think about when 'hanging in the sky' above Heathrow or some other crowded international airport !

2.2 System Types

As a result of the research that has been carried out over the years it is possible to consider most aspects of life in systems terms, which has the distinct advantage of helping us to organise our thoughts about extremely complex situations. The book is mainly concerned with systems where human activity is taking place, and, before moving on to consider in greater detail the properties and characteristics of systems generally, it is worth taking take a brief look at how these relate to other categories.

Precise categorisation is difficult as certain types of systems overlap, depending very much on individual interpretation and point of view. A *social* system, for example, could be regarded as a formally organised set of people with a common purpose (eg the flower-arranging society, the social club), the less formal arrangements that exist within a family, or the relationships that develop between people at the workplace. These frequently occur without any conscious effort, almost by instinct, and could also be categorised alongside other *natural* systems, eg trees, rivers, the universe, etc. Devices that are engineered and built by man are perhaps easier to classify, but even this classification is complicated by considering the role of works of art, music, language, mathematics etc, all of which are essentially abstract, but are contrived and constructed by people.

From this brief discussion it is perhaps easy to understand why the systems debate is an on-going one, and at times essentially philosophical, a factor which will become more apparent as the ideas are used in practice. Nonetheless, from this debate, certain broad categories have emerged which can be summarised as:

a. **Natural Systems** - those that occur *naturally* in the universe, from the lowest minute system form up to the galaxy, including human beings and animals.

b. **Designed Systems** (ie man-made) - such as computers, central heating systems, piston and jet engines, and so on. This category is taken to include abstract system forms, eg mathematics, art, music, philosophy etc.

c. **Social and Cultural Systems** - ie those formed by human beings coming together, either naturally in families, communities, nations, or deliberately in clubs, community centres etc.

d. **Human Activity Systems** - ie systems where human beings are undertaking activities that achieve some purpose. These systems would normally include other types, such as social, man-made and natural systems.

(This summary, illustrated in Fig 2.1, is adapted from that given by Brian Wilson in *Systems: Concepts, Methodologies and Applications*)

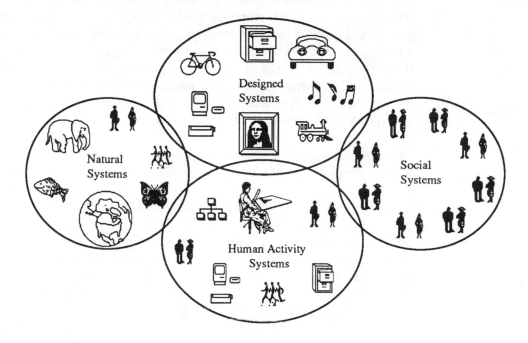

Fig 2.1 - Categories of Systems

This chapter will explore the general characteristics of systems, using a typical man-made device as an example, then cross-relating the ideas to human activity systems in general. This concept (ie of a system where human activity is taking place) may be unfamiliar to many readers, but it is not a difficult one to grasp, and the relationship to human situations and, in particular to formal organisations, will become more obvious as the examination of system characteristics progresses. At the outset, however, it is important to realise that such organisations are seldom engineered on systems lines, but normally structured to group together certain skills or processes, or simply to facilitate administration. A *Human Activity System* is only a concept, and the form of the associated model will depend on the viewpoint selected by the analyst. Nonetheless, the notion that enterprises can be viewed in system terms is a valid and productive one, and provides an excellent basis for both understanding what is happening in practice, or what should be happening to achieve some desired end.

2.3 Basic Systems Ideas

It could be argued that there is no such thing as a system, that it is just a convenient concept that enables us to view a collection of interrelated items as an *ordered arrangement* which, as a whole, achieves some purpose. When considering those that involve human beings this is probably true, as it is not possible to see, hear or touch a social, or political, or industrial relations *system*. However, most of us would think more than twice before paying a large sum of money to a high street retailer or a plumber for an idea, no matter how good that idea might seem. What we expect to get for the money is a set of components (eg pumps, valves, pipes etc), linked together by some means to form a package that has a tangible form, and significantly, when assembled and working together as a whole, will produce the desired end product, such as hot water, central heating, hi-fi sound, and so on. This idea of wholeness or **holistic** properties is one of the basic principles that distinguishes the soft systems approach from other analytical methods, a point that is reinforced by examining most dictionary definitions. For example:

"System - a group of things or parts working together or connected in some way as to form a *whole* " (Collins Standard Reference Dictionary)

Scientists over the years have been typically concerned with breaking whole bodies into their component parts for detailed examination and analysis, but nonetheless, certain ideas have emerged from both scientific and engineering origins that indicate that there is also value in considering how these parts interact to produce the properties of the whole. To understand this process and the implications for human activity in organisations, it is first necessary to consider how a system differs from a simple collection of parts *without* a common identity, and how the process of interaction is achieved and controlled. This can be clarified by exploring the linked concepts of **emergence**, **hierarchy**, **communication**, and **control**.

2.3.1 Emergence

The manner of joining the components together in a man-made system is, of course, not haphazard, but designed to ensure that the end product achieves something new, ie something that did not or could not exist until the necessary connections were made, and would cease to exist if the components parts were separated. **Emergence** or the display of new attributes is easily recognised when considering man-made or designed systems; for example, the ability to provide controlled heating for a house or other building, or high fidelity sound from a number of different sources, and so on. Such properties are more difficult to define when contemplating Human Activity Systems, and will obviously depend on the situation being examined; in broad terms, such a system will have a *collective strength of purpose* as a by-product of the relationships that need to exist in systems terms.

It is interesting to consider the emergent properties that might be exhibited by inanimate objects such as a chair, or a desk, or a house; or even biological systems such as a horse, a frog, or a human being. To illustrate the point, consider the object known as a book, which consists of a number of separate elements, ie chapters made up of a series of paragraphs and diagrams, containing words and printed on paper (Fig 2.2).

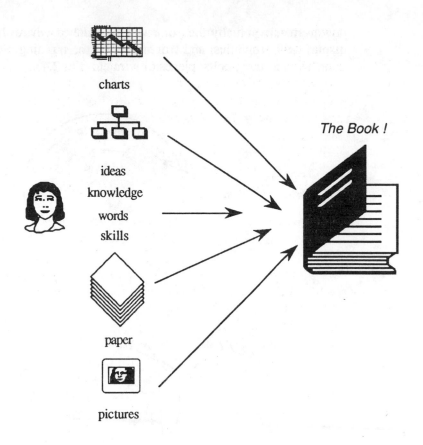

charts

ideas
knowledge
words
skills

paper

pictures

The Book !

Fig 2.2 - The Book Emerges !

These paragraphs, words, diagrams etc, reflect the knowledge and ideas of the author, and the finished product the skills and experience of the printers and publishers. Each of the separate entities exhibit their own characteristics, but it is not until they are bought together that the *book* is formed, displaying different attributes as a whole item than the individual parts displayed in isolation. For example, it provides a collection of ideas in a cohesive form, it is now suitable for inclusion in a library, it has an enhanced credibility, and even the ability to prop up a shaky table !

2.3.2 A Hierarchy of Systems

To misquote John Donne, "no system is an island unto itself", and each will be part of a **hierarchy of systems** (as illustrated in Fig 2.3), with the sub-systems in turn displaying emergent properties. In man-made systems this hierarchy is fairly obvious, eg the microcomputers in a communicating network, or the record, tape and compact disc components of a hi-fi stack etc, each being a *sub-system* of the main assembly, but also systems in their own right, ie a set of components that exhibit new properties when working together as a whole. Extending the hierarchy downwards, eventually the point will be reached where the components will not display new properties, and will simply be the building blocks of the functioning parts, such as nuts, bolts, pieces of wire etc (Fig 2.4).

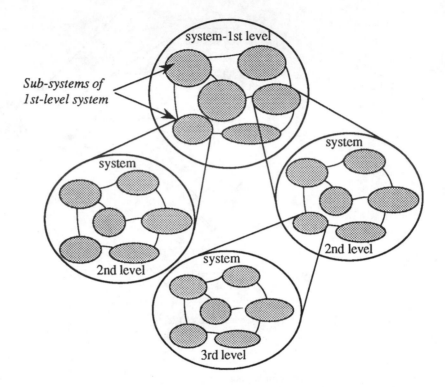

Fig 2.3 - Hierarchy of Systems

Using these relatively simple examples, arguably the system is complete (ie *the highest level of the hierarchy is reached*) when the configuration of sub-systems is such that it achieves the purpose for which the system as a *whole* was designed, eg to heat a building, to produce music, to enable electronic data transmission and so on. In this case it would then be appropriate to consider how it interacts with the surrounding environment. However, the extent of the hierarchy will depend on the perspective being taken, and it is important to realise that any system might be part of a *wider system* that has some controlling influence. We could regard the hi-fi unit as part of an *entertainment system*, the computer network as a component of the *communications system* of an organisation, or both as part of the *cultural system* of the United Kingdom, and so on. At the highest level, it could be one that is too large to examine in any detail and difficult (for us!) to control, such as the world or the universe, etc. The idea that different perspectives lead to different models is explored later in the book, together with examples of how such models can be expanded to reveal the hierarchical structure.

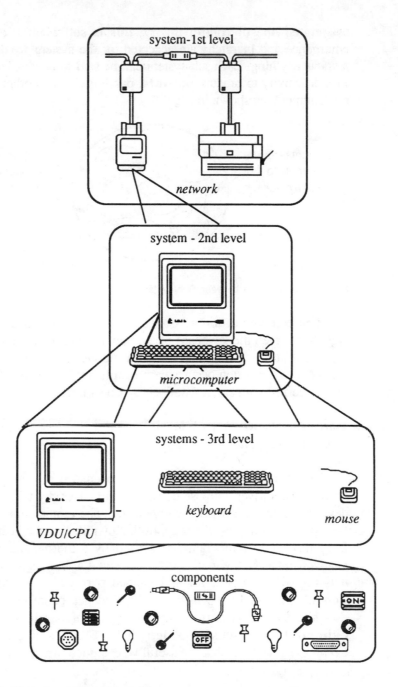

Fig 2.4 - Hierarchy of Man-made Systems

2.3.3 Communication and Control

To function as a whole, it is necessary for the system components to interact in some way, ie there must be some form of communication between them. Without going into the extensive history behind the development of this idea and the associated one of control theory, this is self-evident once the principles of emergence and hierarchy are accepted, or the system would not be able to *do* or *achieve* anything. Each sub-system can be said to receive **inputs**, which stimulate further activity to produce **outputs**, passing either to other sub-systems, or to the environment, as shown in Fig 2.5.

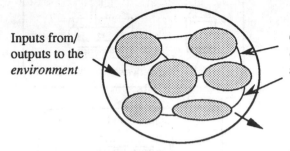

Inputs from/ outputs to the *environment*

Communication links (ie *messages, signals* , etc as *inputs/outputs*)

Fig 2.5 - Interactivity between Sub-systems

The communication messages may be tangible ones, such as the electrical signals between the keyboard of a computer and the operating system, the written or verbal message that pass between individuals or groups, or the less tangible interactions that often occur where people are involved. Many of these messages are concerned with *control,* defined by Checkland as:

"the means by which a whole entity retains its identity and/or performance under changing circumstances"

The definition reflects the need to ensure that a system can continue to accomplish a given purpose despite disturbances, by taking control action once a deviation from preset parameters occurs. The temperature of a hot water system, for example, is regulated and controlled by reducing the heat input when a variance from the preset temperature is sensed by a thermostat. If there is no *control mechanism,* a change in circumstances could affect a system to such an extent that it is no longer suitable for its designed purpose, ie it loses its original identity. Control is normally dependent on the feedback of information or messages about how the system is performing, an idea that has its origins in the theory of **cybernetics**, discussed at length by Flood and Carson in *Dealing with Complexity,* and one that can usefully be applied in practice.

2.4 System Characteristics

Having explored briefly the ideas of emergence, hierarchy, communication and control that underpin systems thinking, it is already possible to see how these can apply to organisational theory. Organisations comprise people undertaking certain activities that (hopefully) contribute to the purpose of the organisation as a whole, sub-sets of activities can be identified with various interactions between them, and control is exercised through decision-making mechanisms, eg the Board of Directors, management, committees, and so on. Ideas can also be developed about measures of performance that could or should exist, the effect of the environment, and other factors that are typical of systems.

It is now appropriate to consider in greater depth the system characteristics which enable these ideas to be of value in practical situations, effectively developing a checklist to help us explore any Human Activity System that we chose to consider. It is easier at the outset to examine a man-made device and consider what characteristics are present that are typical of systems generally, before moving on to apply these ideas to human situations.

2.4.1 The Jet Engine

In the air engineering world there are countless systems, ranging from comparatively simple devices for re-charging an aircraft tyre to sophisticated electro/mechanical configurations that provide cabin heating and pressurisation weapons control, and aircraft navigation aids. One of the most complex to design, build and maintain is the so-called *jet engine*, which has largely superseded piston engines as a device for moving heavier than air machines at sufficient speed to enable them to fly. Conversely, the principles of jet propulsion and the manner of application to a jet engine are generally easier to understand than those of its piston counterpart, and consequently it provides a useful basis for explaining system principles. Put simply, a jet engine draws in air from the atmosphere, then pressurises it using a rotating compressor, which can normally be seen by (carefully) peering down the intake of the engine (Fig 2.6).

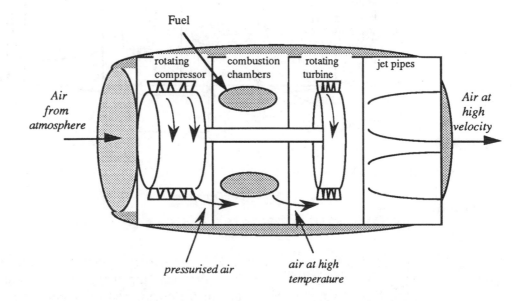

Fig 2.6 - Model of Jet Engine

From the compressor, the air is passed into a number of combustion chambers where it is mixed with atomised fuel and then ignited, producing very high temperatures and, eventually, high velocity air. This is then exhausted to atmosphere via the jet pipe, in the process rotating a turbine, which in turn drives the forward compressor to suck in more air, so that the cycle is repeated and continued.

This is a system in action, and typical characteristics can easily be identified. Firstly, it exemplifies the ideas discussed in earlier paragraphs, ie it comprises a *hierarchy* of systems and components, each with its own properties but combining together through a network of electrical and mechanical devices to produce the whole entity known as a jet engine (Fig 2.7).

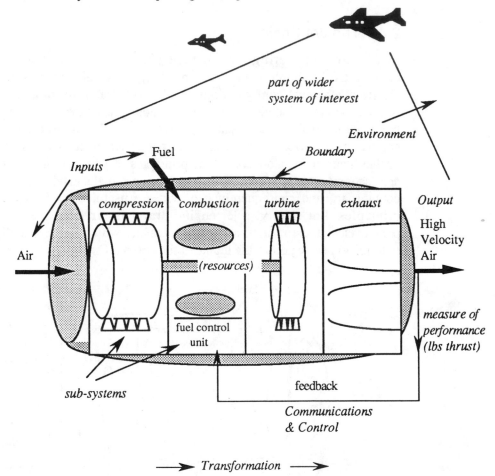

Fig 2.7 - A Thrust Producing System

The *sub-systems*, such as the compressor, combustion, fuel and electrical systems, all interact as necessary, receiving inputs and producing outputs. Messages controlling the status of the engine and its required performance are continuously being transmitted, and control is exercised by sensing the temperature of the escaping exhaust gases, and *feeding-back* regulatory signals that adjust the amount of fuel mixed with the incoming air to keep the power output to that selected by the operator, or within preset safety limits. *Regulation* and *control* are based on the output of the engine (ie pounds of thrust), which is the defined *measure of performance* for this system.

The engine has been designed and constructed with a set purpose in mind, ie to convert an input of atmospheric air to an output of pressurised high velocity air, the difference in velocities producing the thrusting force to move the engine forward and the vehicle (or aircraft) to which it is attached. The idea that a system should have an on-going purpose or mission, and effect a change or *transformation* is a prime characteristic of a system, one that is generally illustrated as shown in Fig 2.8 below.

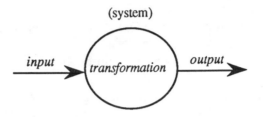

Fig 2.8 - System Transformation

The engine interacts with the surrounding *environment*, being influenced by such factors as ambient air temperatures and velocity; however, it is quite clearly separated from the environment by the recognisable *boundary* formed by the engine casing and panels, outside which the regulating process of the engine itself has no authority or control. Systems which rely on the environment for obtaining inputs and discharging outputs are regarded as **open** systems, by contrast with those that are entirely self-supporting (ie **closed**) such as an astronaut's life-support pack. Within the boundary, the system has *resources* for its own use, in the form of the static (unchanging) sub-systems and components, and the variable resources of fuel and air. The interaction with the environment, the existence of a separating boundary, and the availability of resources for the system's own use are further characteristics of systems. When mounted in an aircraft or other vehicle, it becomes part of a *wider system of interest*, from which it receives control messages adjusting performance as required.

2.4.2 Summary of System Characteristics

The jet engine description highlights certain characteristics which could be considered typical of those configurations referred to as systems. In the case of the jet engine and other man-made constructions, if any of these were not present, then then the device could no longer function as a *whole* entity. These essential characteristics, more accurately described as concepts in relation to human activity, provide a useful checklist for ensuring that *Human Activity System models* are complete, ie all the necessary factors have been considered by the analyst when they are being developed. These are summarised by Checkland in an intellectual construction known as the **formal systems model** (summarised in Fig 2.9), which can be used as the basis for examining models of human activity to ensure that they are well-formulated, and satisfy the following criteria:

- The system represented by the model has an ongoing purpose, ie it exists for a reason, and achieves some transformation or change
- There are measures of performance so that it can be shown to be effective, and which can be used as a basis for measuring efficiency
- There is some mechanism for control or regulation, and a decision-making process
- It has components that are themselves systems
- It has components that interact
- It exists as part of a wider system or systems, in an environment with which it interacts
- It has a boundary which encloses the area that the regulating mechanism has under control
- It has resources for its own use, under the control of the regulating mechanism
- It has some expectation of continuity, and can be expected to recover from disturbances

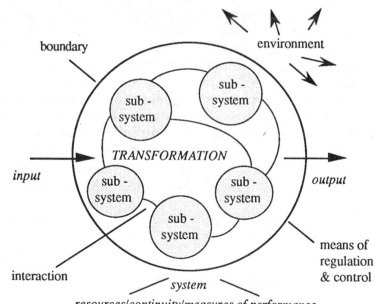

Fig 2.9 - Summary of Typical System Characteristics

The importance and practical value of the formal model will be realised as Human Activity Systems are discussed in later paragraphs, and when considering the case studies described in Chapters 10 to 14. Human Activity Systems can be regarded as *open* systems, as there is a continual interaction with, and a reliance on, the surrounding environment. Unlike man-made constructions, however, these have no tangible form, and can be viewed from a number of different perspectives, each of which would result in a different model. Whereas the jet engine can be described unambiguously in terms of its weight, shape, form, power output and so on, no similar consistent description can be derived for a system of human activity. (The notions of viewpoints, perspectives and associated models is examined further in Chapter 6.)

Furthermore, whereas the sub-systems of the jet-engine are readily identifiable as physical components, these are generally in the form of activities when considering organisations, some of which may be observable, but others taking place as mental processes. Similar problems arise when considering other factors, such as communications, boundaries, decision and control mechanisms, the interaction with the environment, and measures of performance. Nonetheless, the exercise of examining human activity with these in mind is both revealing and productive.

Brian Wilson also points out that, although a Human Activity System is usually modelled as a series of activities, it should not be forgotten that an accompanying social system is also being defined, which will have a strong bearing on whether or not certain changes will be accepted. To emphasise the point, he illustrates the relationship between the two types of systems in a manner similar to that shown in Fig 2.10.

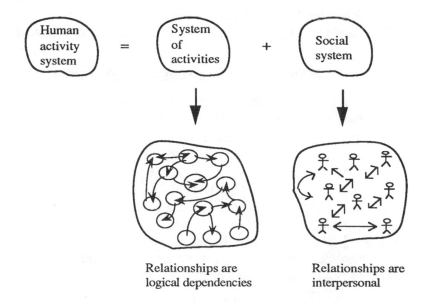

Fig 2.10 - Systems within Human Activity Systems

These ideas are incorporated in the Soft Systems Methodology which forms a large part of the later discussion. When applying this methodology, models are constructed from a variety of viewpoints, and the relationships between the *people* involved in organisational activities are explicitly considered.

2.5 Exploring Human Activity Systems

To summarise the points made so far, it is worthwhile to discuss them in relation to a typical enterprise, one that could be regarded as a Human Activity System. Based on the idea that this is a system where:

"human beings are undertaking activities that achieve some purpose....."

we can see that practically all areas of human endeavour could be regarded in these terms, especially where people are grouped together to satisfy some business need (eg manufacturing a product, providing a service, banking etc). It is worth recalling that system models seldom reflect organisation structures or relate to functional groupings, as, depending on the viewpoint taken, systemic activities may permeate throughout the organisation as a whole. For example, if we are concerned with identifying what is needed to operate an effective industrial relations system, activities such as *negotiating, consulting, assessing staff/management needs* come to mind. However, these will not be confined to a particular functional group or appear as part of the formal structure, but will be taking place as a series of lower-order actions in various parts of the organisation. Bearing this in mind, it is nonetheless revealing to explore a typical enterprise to determine how system characteristics are identifiable in practical situations, which also helps to put the points made in this chapter into context.

2.5.1 Management Services as a System

The term *management services* is a generic one that covers a multitude of activities (and sins?) carried out in support of organisations. The services provided can include advice on office equipment and layouts, market appraisal, operational research, and the study of working practices and methods. The most typical arrangement of disciplines in recent years is one that has combined Work Study and Organisation and Methods (O&M) personnel with specialist advisers on the use of Information Technology in the office environment, to form a Management Services Unit (MSU).

Examination of this typical formation, as shown in Fig 2.11, enables us to consider how systems characteristics might be identified in real situations, and also highlights some of the difficulties that can occur when attempting to define and identify them in practice.

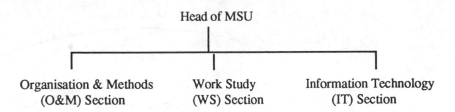

Fig 2.11 - Typical MSU Structure

In systems terms (Fig 2.12), a MSU has an identity of its own and exhibits properties of a whole body, properties that would not exist if the specialist functions were separated. It has a collective strength of purpose, and, instead of reflecting the skill limitations of the separate disciplines, presents a more up-to-date image of a pool of problem-solvers for management to make use of wherever necessary. These emergent properties are represented by the establishment of a post to head the MSU, someone who will oversee and co-ordinate the work of the unit as a whole, and can speak on its behalf at management/board meetings etc.

Whilst recognising that functions and systems are not synonymous, the activities of the MSU can be considered as contributing to a larger system or set of systems that form the company, local authority, government body and so on. Organisationally, the Unit could be part of a department (eg Personnel, Production etc). The unit as a whole will aim to achieve a transformation, which in broad terms could be expressed as:

'to gain improvements in the efficiency/effectiveness of the ...Company/Local Authority etc...'

At this level of definition, we could say that the input is a *desire for improvement*, which is transformed by the efforts of the management services practitioners into an output of *desire satisfied*. At a more detailed level of resolution, the inputs could be formal requests to review staffing levels, improve procedures, reduce costs etc, and the outputs tangible items such as reports, presentations, memos and so on; which are the means of achieving the desired improvements in terms of efficiency and/or effectiveness.

The individual sections would undertake sub-sets of specialised activities as a contribution to the fulfilment of the main transformation, such as improving the cost-effectiveness of groups of manual workers, reviewing office procedures and cost-effectiveness of clerical staff, and ensuring value-for-money in the installation of computer support systems, taken in this context as being responsibilities of the WS, O&M and IT sections respectively. When considered in this way, a hierarchy of systems is evident, with communication between the MSU sub-systems on matters of mutual interest (eg staffing issues related to information technology developments), and an interaction with other systems in the department and the company.

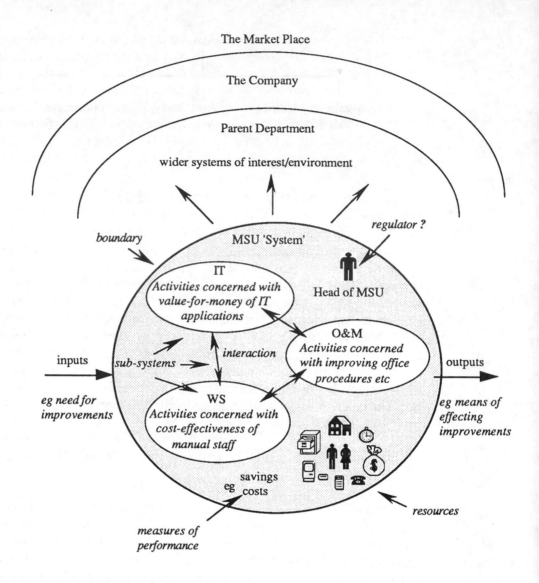

Fig 2.12 - The MSU as a System

Consider next the requirement to have a measure of performance. Data about how the practitioners use their time is usually recorded for charge-out purposes, and, taking into account salaries and other costs, this can be used to assess input (to studies etc) against output in terms of staff savings or increases in profits, etc. This simple equation could be misleading, of course, as it wouldn't take into account qualitative improvements which cannot always be measured in tangible terms. Other indicators of success might be more appropriate, such as the percentage of recommendations accepted and implemented by clients, and so on.

Formally, the output would be reviewed at periodic intervals by some management committee or forum, acting in this respect as regulators who could terminate the existence of the unit if the service it provided was not considered worthwhile. On a day-to-day basis, the decision taking process could be delegated to the Head of the MSU, who would be responsible for overseeing the work of individuals and controlling the general direction taken, ie monitoring performance so that corrections can be made if necessary. Taking this lower level view (ie that the regulator is the immediate boss) makes it easier to define a system boundary, ie encompassing the area that the Head of the MSU has under direct control, and the resources at his/her disposal; the human resources, equipment such as stopwatches, computers, and the accommodation the unit occupies. It is always debatable who really owns and controls these resources; it could be the company, the landlord, the Head of the MSU, or even the client; which serves to illustrate the difficulty that can arise when dealing with human situations rather than man-made systems!

Similarly, there can be problems when attempting to specify the environment of the system being considered, which obviously depends on how the boundary is defined, taking account of the wider systems that interact with groups such as the MSU. In practical terms, a useful rule-of-thumb is to determine whether or not the relationship between the unit being examined and other elements of the organisation is one of control, or simply an interchange of inputs, outputs and influence. If *control* is exercised from outside, then the unit should be considered as part of a wider system of interest; if it only receives inputs, provides outputs, and is *influenced* by other organisational elements, then they are part of the surrounding environment. (This argument is developed more thoroughly, together with other useful rules on boundary definition, by Flood and Carson in *Dealing with Complexity*.) Boundary definition will depend on the circumstances of each organisation, in particular the degree of autonomy of a sub-section and the authority delegated to the person in charge.

In terms of continuity, the MSU has probably existed for some time. However, as MS activities do not always contribute directly to the output of the organisation as a whole, continued existence is not guaranteed, and will be subject to review in light of changing circumstances. It will only recover from disturbances if it can continue to be of value in those new circumstances; a good example is the ability in recent years to develop and sustain the necessary skills to provide advice about new technology. It is important, therefore, that a supportive system such as a MSU includes some mechanism for forecasting future demands and taking the necessary action to ensure the unit adapts to meet them.

It is easy to appreciate from this examination of just one enterprise that nothing is as clear cut, precise or predictable as it is when considering an engineered system such as a jet engine. Different viewpoints would interpret the situation in different ways; boundary definition is also a matter of interpretation, and there are a number of hierarchical levels to be taken into account when deciding ownership of resources, control and regulating mechanisms. However, the Soft Systems Methodology includes guidelines for reviewing these factors when applying systems ideas in practice, providing a framework for the analysts to structure their thoughts about these and many other aspects of complex human situations.

2.6 Summary

This chapter aims to set the scene for those that follow, drawing out from various examples the ideas that underpin systems thinking, and exploring the concept of a *Human Activity System*. As a preliminary exercise in soft systems thinking, a hypothetical management services unit has been examined to distinguish system characteristics, highlighting the fact that, although they can be identified even by a cursory examination, this still leaves many questions unanswered. The next chapter examines some of these questions, and also considers how the approach can be used to aid the process of understanding and improving organisations.

It worth noting that the use of systems ideas when exploring organisations is not unique to the soft systems approach, but has featured in many research projects primarily concerned with organisation and management theory. It is not the purpose of this book to examine these in any depth, but, for readers who wish to gain an improved understanding, the table in Fig 2.13 provides a convenient summary of the early contributors to the debate.

DATE	THEORY	RESEARCHERS
1951	Socio-technical systems	Trist & Bamforth
1958	Open systems/work design	A.K.Rice
1961	Mechanistic/Organic management systems	
	Environment and structure	Burns & Stalker
1965	Technology and structure	Woodward
1965	Types of environments	Emery & Trist
1966	Systems approach to organisations	Katz & Kahn
1967	Environment and structure	
	Contingency theory of organizations	Lawrence & Lorsch
1968/9	Environment, technology and structure - multi-dimensional approach	Pugh, Hickson and others

Fig 2.13 - Summary of Early Systems Theories/Research

3 The Use of Soft Systems Analysis

3.1 Introduction

Even from the few examples discussed so far, the reader may appreciate some of the attraction of soft systems thinking; it provides an interesting and new way of viewing human activity, and it has an intellectual depth that is both stimulating and rewarding. The exercise of using the approach can be satisfying in its own right, but its value when addressing real-life problems is not always so clear. Many analysts who have some knowledge of it find difficulty in understanding where it can be used, and what benefits can be gained by doing so. It is perhaps unfortunate that the term system is used in this context, as, although it is obviously valid, it invites comparison with the better-known systems analysis *techniques*. This can lead to misconceptions about the style of the approach, and false expectations that the outcomes will always be measurable in terms of savings or other gains, or clearly recognisable as computer systems specifications, and so on. Soft systems thinking is not solution-oriented, but biased towards clarifying the problems that are felt to exist in given situations. Once this has been achieved, other analytical techniques may have to be applied before solutions are found. Where the emphasis is on end results, the soft systems approach is sometimes regarded as an interesting but time-consuming diversion from the customary or *proper* way of doing things.

It is easy to dismiss this scepticism on the grounds that analysts who take this view are conditioned to means-end thinking only, but any examination that consumes the time and resources of an organisation must justify itself in some way in terms of the benefits that it brings, even if these are the less tangible ones of increased knowledge or understanding. So questions such as *"Where should or could the soft systems approach be used ?"* and *"What value or benefits does it bring ?"* are relevant ones, and must be answered satisfactorily before it will be more widely accepted by the problem-solving professions.

A comprehensive summary of typical situations where the approach can be used to good effect is given in Chapter 9, with more examples in those chapters that cover specific practical applications. To put the approach into context, we are concerned here with addressing the questions in broad terms only.

3.2 Systems Analysis - The Hard/Soft Division

The terms **hard** and **soft** are used frequently in most explanations of the soft systems approach, and, before examining this in any detail, it is first necessary to make clear the distinction between the two. The terms are essentially comparative ones, and are used to distinguish between methods of examination that address clearly defined problems (ie those that are suitable for the application of prescriptive *techniques*), and others that are used when the problem is not clear at the outset, and a preliminary investigation is required to *identify* and *select* the problems to be solved. The latter type of examination applies to situations that are regarded as *unstructured* or *soft*, inevitably involving people working as individuals or groups towards some common organisational or other goals. To clarify the difference between the two approaches, consider Checkland's explanation of the development of *hard* systems analysis, as summarised in the following section.

3.2.1 The Development of Systems Analysis

In *Systems Thinking, Systems Practice*, Checkland explains that systems analysis originated at the latter end of the second world war, with the emergence of a project known as Project RAND (ie Research and Development). This was initially a collaborative venture between the American War Department and Douglas Aircraft company for a "study of intercontinental warfare, other than surface", with the object of advising the Army Air Forces on devices and techniques. Eventually, because of some conflicts of interest, such as the difficulty of avoiding favouritism when awarding contracts, the RAND Corporation broke away from Douglas and became a non-profit making concern, carrying out long-range research activities as an aid to strategic and technical planning and operations.

The approach to problems was one that had developed from the involvement of scientifically-trained civilians with military operations during the second world war, carrying out what is now referred to as Operational Research. In broad terms, this consisted of :

a. Deciding what needs to be done, stating clearly the aim or objective.

b. Determining alternative ways of achieving the objective.

c. Appraising the costs of each alternative.

d. Building a model of the different alternatives, usually a mathematical one, and testing each model under different conditions.

e. Deciding, on the basis of pre-defined criteria, the preferred or optimal alternative.

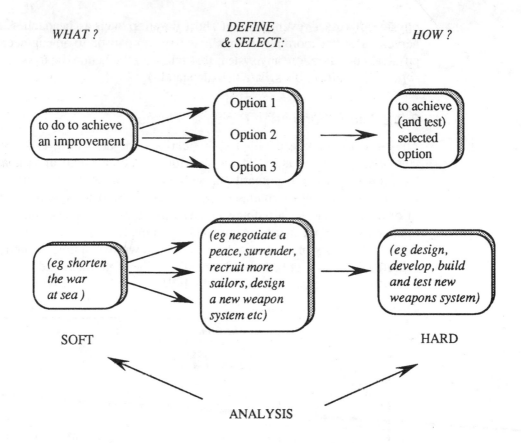

WHAT ?

DEFINE & SELECT:

HOW ?

to do to achieve an improvement

Option 1

Option 2

Option 3

to achieve (and test) selected option

(eg shorten the war at sea)

(eg negotiate a peace, surrender, recruit more sailors, design a new weapon system etc)

(eg design, develop, build and test new weapons system)

SOFT

HARD

ANALYSIS

Fig 3.1 - The Soft/Hard Division

The most significant aspect was the emphasis on *how* to solve problems that were clearly defined at the outset by the policy makers or military advisers. For example, how to provide a new weapon system to achieve a 50% increase in the success rate of air to surface missiles. Before the stage of stating problems clearly could be reached, the broader questions of *what* to do to shorten the war at sea, eg negotiate a peace, surrender, design a weapon system etc, had been posed and answered at a higher level (Fig 3.1).   ) copy

Hard systems thinking and analysis is essentially concerned with the question of *how* to achieve a predetermined aim; **soft** is concerned with defining the options for improvement, in other words, addressing the *what to do* question. It is also essentially committed to the examination of human activity, which is the other soft part of the equation. The behaviour of humans is largely unstructured, and, although certain tasks may be carried out to a prescribed set of procedures, an individual will seldom perform them in exactly the same manner or at the same speed. (These differences in behaviour are reflected during work measurement exercises by *rating* the performance of individuals to even out fluctuations in performance.) The relationships between individuals also varies, and attitudes can significantly affect the efficiency of human activity; we are affected by mental and physical illnesses, even those of short duration such as headaches, hangovers, nerves, 'Monday morning feelings', etc. By comparison to an engineered or *hard* physical configuration, any system that relies on the human input is amorphous or *soft*. (Although, at times, hard to understand !)

3.2.2 The Difference in Practical Terms

The need to prepare detailed specifications for designing and programming computer systems has led to the development of a range of systems analysis techniques. These are primarily concerned with breaking each problem into its component parts, representing activities that the computer system must undertake in order to carry out a defined process, together with the associated data and relationships. Work measurement techniques evolved from similar origins, ie the concept that work processes could be broken down into simple elements, ones that could be clearly described and observed, and therefore measured, ie the principles of *scientific management* formalised by F.W Taylor (1856 - 1915).

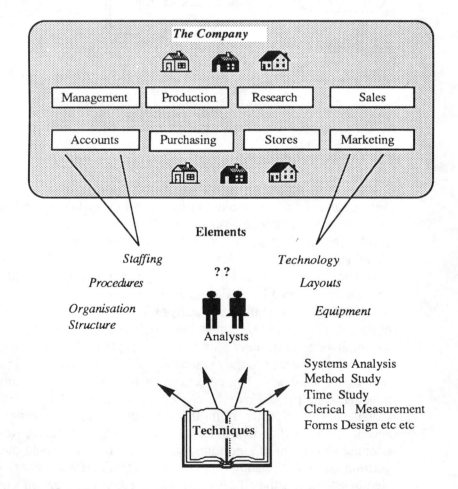

Fig 3.2 - Typical Hard Analysis Approach

In each case, a part of the organisation is selected and then scrutinised in detail using the appropriate techniques. Depending on the type of review, an examination of the structure, levels of staff, office procedures and so on will be undertaken, concentrating only on parts or *elements* of the subject organisation (Fig 3.2); significantly, those elements will have a tangible form, unlike the systems that might be considered. In these circumstances, it is reasonable to assume that the problem is a *hard* one, ie it has been defined for the analyst and is suitable for the application of a technique that gives a predictable type of solution, such as the design and construction of a computer system, improvements to methods of working or the layout of workplaces, or a reduction in the time taken to complete a task, and so on.

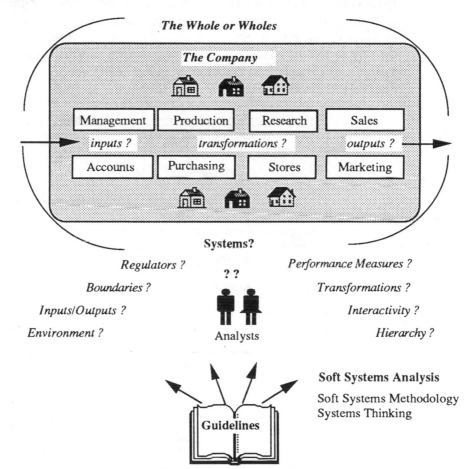

Fig 3.3 - Looking for Systems

The soft systems approach on the other hand is concerned with clarifying what the problems are in any given circumstances, as a step towards making some improvement. To accomplish this, instead of addressing selected elements, a *holistic* view is taken by considering the organisation as a system or series of systems (Fig 3.3).

3.2.3 Systems and Organisations

Using this approach, ideas are developed about the activities that are needed to achieve defined transformations, posing further questions about systems characteristics, eg measures of performance, communications, control, and so on. However, relating systems ideas to organisations can be difficult, a point reinforced in the previous chapter when discussing a Management Services Unit in systems terms. Although it is sometimes useful to consider such groups of resources as systems, in this respect the term is being applied loosely, and on closer examination it is unlikely that the requirements of the formal systems model would be fully satisfied. *Systems* and *functional groupings* are seldom synonymous, and this is where further complications arise. Within any enterprise there may exist sets of activities, not necessarily arranged together, but which could be said to achieve some purpose overall. This purpose may be defined and related directly to the organisational objectives, or to a supportive role (eg a *financial* system), or to undefined personal or group objectives. A summary of types of systems that could be relevant to an organisation is given in Fig 3.4.

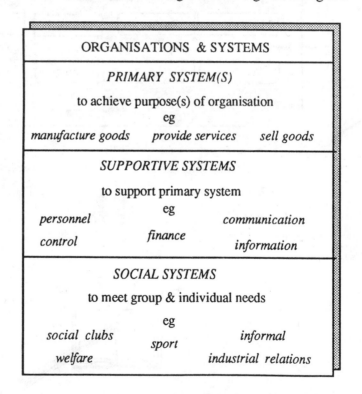

Fig 3.4 - Examples of Systems in Organisations

Confusion arises because these systems are not observable, or apparent from any examination of the organisation structure charts, ie they are not recognisable as tangible configurations, but can be envisaged as a set of interrelated activities. Nonetheless, it is relevant to consider them when attempting to understand what is actually happening in practice, or what should be happening in order to achieve some desired end. The systems could include those that are concerned with achieving the primary purpose of the enterprise, those that provide support in some way, and those that come under the broad heading of social systems. Some of these latter ones may have a tangible *basis*, eg social and sports clubs, welfare organisations, whereas others are essentially conceptual relationships, such as the industrial relations system referred to earlier.

To further confuse matters, it is also possible to consider human activity in more abstract terms, and develop system models from somewhat contentious viewpoints, eg systems to *happily employ people,* or to *make enormous profits for shareholders,* and so on. These *issue-based* systems are discussed at length in Chapter 6.

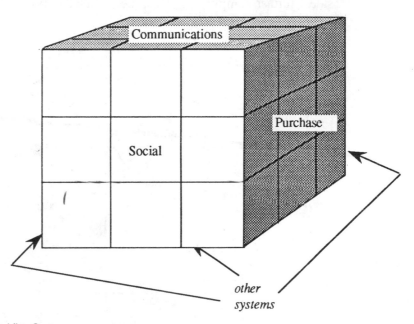

Fig 3.5 - Rubik's Systems

Staying with the more obvious examples for the moment, to clarify the notion of nested systems in an organisation, an analogy can be made with a Rubik's cube that is constructed from a series of systems, for instance financial, social, communication, information, purchase, and production systems. Each of these contributes in some way to the achievement of the organisational goals, such as producing motor cars, providing local authority services and so on; the operation of the system (or the cube) as a whole is largely dependent on the effectiveness of the component sub-systems or activities. Theoretically (and ideally from the point of view of the analyst), all *related* activities would be grouped together and appear on a single face of the cube, so that they could be seen and examined without any difficulty (Fig 3.5).

However, few if any of these would ever appear as an identifiable group of activities within an organisation, but are scattered throughout, with elements of each found within the functional groupings of departments, sections, branches etc, in effect like the cube in its uncorrected form (Fig 3.6). Whereas there are physical links holding the sub-sets of the cube together, such links may not exist in any tangible form in an organisation. For example, a *financial* system may well have a specific base such as the accounts or treasurer's department, but throughout the organisation there will be many activities going on that could be considered as part of this system, eg paying out money, calculating costs, using up or buying in resources, and so on. In some cases there may be discernible links between these activities in the form of information flows or the movement of resources and so on; in other cases the connections will be *conceptual* ones as envisaged by the analyst. Additionally, considering the number of perspectives that could be relevant to any organisation, it is possible to imagine far more systems than the Rubik's cube has sides, further complicating the process of identifying and improving them when required.

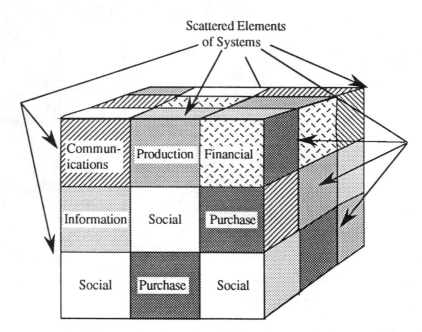

Fig 3.6 - Nested or Scattered Systems

One further point needs clarifying; the analyst who uses the soft systems approach is not necessarily concerned with reforming the organisation on systems lines, there are many reasons why this would not be viable. However, the exercise of thinking about the situation in these terms leads to a new understanding of what could or should be happening *below the surface of the real world*, why certain problems occur, and, as a consequence, how improvements can be made.

3.3 The Advantages of Soft Systems Analysis

To summarise, *hard systems analysis* addresses those parts of an enterprise that have a tangible form, eg the structure, the levels of staff, the equipment they use, the accommodation taken up etc. Soft systems thinking, however, considers the systems that could be envisaged throughout, and, in particular, those that involve human activity. In general terms, therefore, *soft systems analysis* can be defined as:

'The use of systems ideas to analyse (ie examine) soft situations to identify where problems could exist'

It naturally follows that, once the problems have been clarified, it will be necessary to utilise appropriate hard techniques to solve them. The soft systems approach to problem solving, therefore, supplements rather than replaces other forms of analysis. Frequently it is used as a means of improving the analyst's understanding of a situation in the early stages of an investigation, and at subsequent stages to clarify where changes might be beneficial. In this respect, it is an aid to the analyst in making sense of situations that are inevitably complex and *cluttered* by people and their individual perspectives, political influences etc, and many other factors that are relevant but difficult to view in any structured or methodical manner. Fig 3.7 summarises the use of systems ideas to explore a human situation, focus on a particular area where improvements seem possible, and apply the appropriate techniques for identifying savings, improving methods etc.

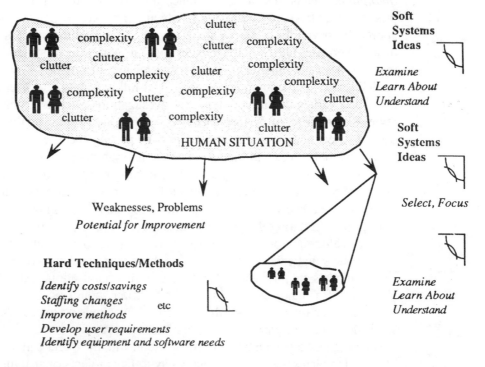

3.7 - The Hard/Soft Relationship

Assuming that no human situation is perfect, the answer to the question *"where should or could the soft systems approach be used"*, therefore, is *anywhere*, or anywhere there is a need to enhance an observer's understanding so that improvements can be considered. The broader question of what benefits the soft approach can bring should now be easier to answer, ie it leads to increased understanding and helps to clarify potential areas of weaknesses. Significantly, the approach encourages an analyst to take an overview of situations and consider relationships that may not be apparent when using *hard* techniques, and this is of particular value in today's climate. It is concerned with the *whole*, and helps to ensure that improvements to one part of an organisation are made in full recognition of all the influencing circumstances. How this is achieved will be clarified later in the book, but in brief, it requires a summary of all relevant factors to be made at the outset of a study and the development of system models to illustrate the activities that are *logically* necessary to accomplish a defined end, disregarding, in the first instance, the structures that actually exist. As a consequence, it crosses traditional boundaries and encourages ideas about wider relationships and how they can be strengthened or sustained.

Furthermore, regardless of the business or other aims of organisations, few if any can function effectively unless some characteristics of systems are present in their day-to-day operations. *Measures of performance* and mechanisms for *feedback and control* in particular are essential to an organisation's survival; frequently these aspects are not formally recognised in any way, or consciously considered when a review of activities etc is being undertaken. In many situations *quantifiable* measures of performance are difficult to define; nonetheless, the process of addressing such matters, which is an integral part of the soft systems approach, inevitably produces some means of judging the success (or otherwise) of an enterprise. Other characteristics that are desirable in systems terms (eg the resource needs, communication links and related information flows etc), will also be deliberated as the approach is applied. It follows that soft systems ideas can be used for a variety of different types of investigations, such as general departmental reviews or reviews of performance, value-for-money audits, analysis of functional needs to meet new commitments etc; and, by putting specific developments into an organisational context, in support of computer systems analysis.

It will later become apparent that, to progress the modelling exercises to a satisfactory and credible conclusion, a fair degree of effort and time is required. Apart from the knowledge gained during these exercises, it may seem that they are only of benefit during the period of the study; however, experience has shown that well developed models not only provide a deep insight into the complex relationships that exist at the time of analysis but can also have a lasting value, ie certain models are *fundamental* to each specific organisation, and will only become irrelevant if a major change in role or function should occur. This realisation is significant for those analysts that are either full-time internal consultants, or deal with the same companies or corporations on a regular basis. Careful, dedicated development of agreed models in consultation with the client can result in a number of basic templates that underpin many day-to-day projects or analytical activities. In particular, the systematic derivation of an information database for an organisation (see Chapter 11) will provide a long-standing framework for all manual or technological advances related to the collection, collation and use of the information.

The process of soft systems analysis, therefore, is not just relevant to a particular project or of isolated value, as the models derived can remain extant for many years. Furthermore, with the exception of models that are essentially related to the primary tasks or roles, very few are unique. For example, the logical set of activities that are needed to ensure good industrial relations within one organisation apply equally to others, and a series of *typical models* or templates could eventually be constructed. (This is essentially the same principle used in the FAOR package, as discussed in Chapter 14, for the *Generic Office Frame of Reference*.)

3.3.1 Thinking Logically

The discussion so far has been centred around the use of the soft systems approach to address broad-based organisational issues. The ideas normally applied as part of a packaged methodology can also be used in isolation to aid creative and logical thinking, *systems thinking* and *logical thinking* being almost synonymous in this respect. Some examples of how this can assist with normal studywork are summarised below, with further details given in Chapter 9.

Carrying Out a Study

For example, at the outset of an examination a model can be developed showing an ordered or *logical* sequence of activities needed for the examination to be successful, ie summarising how to undertake a project in a *systematic* manner. Once the study starts, the analyst is normally bombarded with a series of visual and aural images, appearing as a clutter of apparently unrelated facts which do not immediately make a lot of sense. Systems models can provide the analyst with a template to expedite the process of *orientation* in strange circumstances, albeit a template that may have to be modified as the study progresses.

Putting into Context

Even if the brief is limited to an examination of a discrete functional group within an organisation, the exercise of developing a model that illustrates the activities needed to achieve the purpose of the function, showing also how it relates to other parts of the organisation, will enable the operation to be viewed in a wider context, giving rise to more acceptable solutions. To illustrate this point, if the brief was to replace manual equipment in the typing pool with wordprocessors, the process of typing could simply be regarded as one of *converting manuscript into printed text*. In a wider context, the quality, style and speed of output is related to the needs of the organisation as whole, which may be concerned with image, customer relationships, and so on, all factors which have a bearing on the final choice of equipment.

Viewing typing or any other discrete groups of activities in systems terms will encourage the *context* of the activities to be considered during the development of improvements of any kind (Fig 3.8).

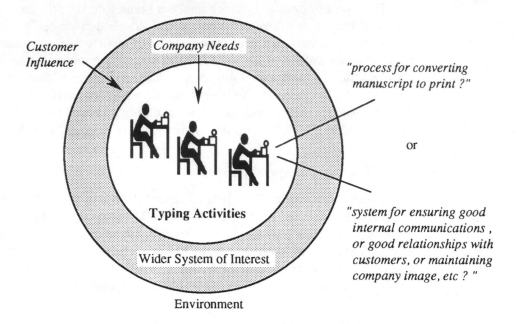

Customer Influence

Company Needs

"process for converting manuscript to print ?"

or

Typing Activities

Wider System of Interest

"system for ensuring good internal communications , or good relationships with customers, or maintaining company image, etc ? "

Environment

Fig 3.8 - Putting into Context

3.4 Summary

Up to this point, the arguments in favour of using the approach, and the advantages that it brings have been explained in broad terms, as summarised in Fig 3.8. In many respects the extent and style of application will depend on the ability of the analyst to look beyond the confines of specific analytical disciplines and make use of systems principles in whatever way is appropriate in each set of circumstances. The following chapters will explore ways of applying these principles in practice, first by examining the Soft Systems Methodology, and then by introducing more specific examples of how this Methodology and system ideas generally have been used to good effect.

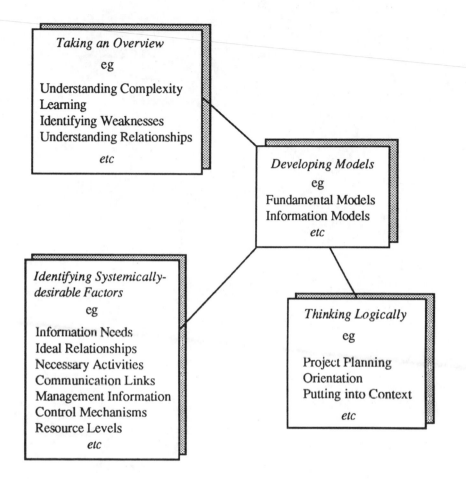

Fig 3.9 - Summary of the Use of Soft Systems Analysis

4 Introduction to the SSM

4.1 Introduction

The preceding chapters have covered briefly the theory of systems, and the concept of a Human Activity System which is fundamental to the particular application known as the Soft Systems Methodology. This approach, referred to from here on as the SSM, and much of the previous discussion about systems concepts, is based on the work of the Department of Systems at the University of Lancaster, and its associated consultancy company ISCOL Ltd. The SSM was developed and tested throughout a large number of *Action Research* projects, described by Brian Wilson as:

" *...simultaneously bringing about change in the project situation (the action) while learning from the process of deriving the change (the research)*"

In effect, this consisted of investigating problems in organisations outside the University sphere whilst developing ways of addressing these problems, and describing them in systems terms. As a result of these projects, the SSM received a thorough evaluation in real-life situations during the development phases. It is now used in its own right for examining organisations with a view to making improvements, and also features in a number of methodological packages, including many that are concerned with deriving the user requirements for computer systems; notably Multiview (*Information Systems Definition: The Multiview Approach* - Wood-Harper, Antill, and Avison, Blackwell Scientific Publications 1985), COMPACT, developed by the Civil Service (Crown Copyright, 1986), and the FAOR package, which is described further on in this book. In each case, the SSM is utilised primarily to gain an understanding of an organisation as part of the process of determining where new technology could be used to good effect. This chapter is concerned with taking an overview of the SSM before exploring each stage in detail as the basis for the practical applications that follow. For convenience it will be described sequentially; the point is often made that it isn't normally, or necessarily, applied this way, a point that soon becomes self-evident in practice.

The SSM is not a technique, ie a method that requires certain procedures to be followed in order to obtain a predictable outcome, but is a set of guidelines for applying systems ideas to problem situations. Although these guidelines help an analyst approach investigations methodically, they still allow considerable scope for individual interpretation. The examples given in later chapters will demonstrate that, in the real world of consultancy and analysis work, investigations using the SSM seldom proceed on any prescribed lines, and it is also necessary at times to make some compromises within the general framework so that a study progresses to the client's satisfaction.

Having said that, this book is not about compromises as a rule, but about recognising that it is sometimes necessary to put dogma to one side and make use of elements of the systems approach to produce results without aiming for perfection. It is a matter of perspective; practitioners must obtain results that are meaningful in their terms, such as savings to the client, improved organisation structures, suitable computer systems and so on. My advice to would-be exponents is, if your particular application of this or any methodology works for you, then it must be considered acceptable.

I have never found the SSM easy to use, or, in the first instance, easy to understand, and the learning curve is a long one for both the theory and the practice. Recognising these difficulties, I have endeavoured to simplify the explanations wherever possible, and give them a personal interpretation that has been found relevant in practice. Hopefully, this will make the learning process easier for newcomers, particularly by considering problems that have arisen, together with a few 'tricks of the trade' which may be of value when first using the approach.

The following paragraphs give a brief description of the SSM overall, to set the scene for the more detailed explanations of each stage covered in later chapters. As the description of the SSM and its use unfolds, advice is given on particular aspects that seem, from experience, to cause difficulties for professional analysts, in the hope that this will not only ease the path of existing users, but encourage more problem-solvers to make use of the methodology. There is a bonus to this; the client quite often tends to see the analyst in a different light after hearing the results of a soft systems investigation. Perhaps this is because of the new insight to the organisation that is revealed, or that it makes a nice change from resurrecting old problems with typical solutions; it could even be due to a touch of the *emperor's new clothes* syndrome, ie the client doesn't wish to appear foolish by not seeing matters in the way the analyst does. This, of course, would seem grossly unfair to the clients !

4.2 The Soft Systems Methodology - An Overview

The action research projects which were the basis for the SSM used the principles of hard systems analysis to investigate soft situations, and from these experiences the SSM model was formulated, showing diagrammatically the seven stages of the methodology (Fig 4.1). This model is rather forbidding at first glance, showing as it does a series of actions described in terms that are probably unfamiliar to many people. However, it quickly becomes apparent in practice that many of the actions needed to complete each phase are similar to those carried out in most investigations, regardless of the method or approach used.

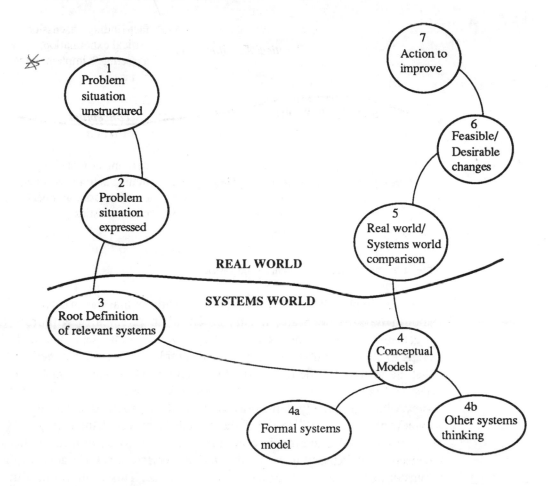

Fig 4.1 - The SSM Model

The diagram is essentially illustrating a set of guidelines, and each stage is described using high-level generic terms (eg *Problem Situation Unstructured, Real World/Systems World Comparison*, and so on). In practice, the analyst will be interviewing, observing, reading-up on background material, and, during the later stages, deciding in conjunction with the client acceptable ways of making changes, and how these can be implemented. A line is drawn on the diagram between such 'real-world activities', and those carried out in the 'systems world', where the analyst withdraws from the examination of the real situation, and objectively considers relevant system models (Fig 4.2).

A pictorial summary of the actual situation is prepared as a **rich picture** during the process of information gathering, a technique which may be unfamiliar to many readers. However, it is the specific inclusion of the systems thinking stages, shown beneath the line, that distinguishes this approach from others.

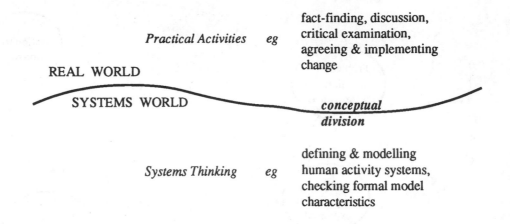

Fig 4.2 - The 'Real-world/Systems-world' Division

The implication of this conceptual division is that the two types of activities should be kept separate; it isn't essential to physically withdraw from the organisation being studied during the systems thinking stage, although this often helps to clear the mind. However, it is desirable that a clear distinction is made so that the facts about the real world do not unduly influence the analyst at this time. The system models which are developed, taking account of a number of relevant viewpoints, are clearly defined as the modelling progresses, and are then used to explore the real world to see if the system is reflected there (Fig 4.3). In other words, the real situation is critically examined to determine if the activities considered necessary to *make the defined system work* (ie achieve the desired transformation) are actually going on in practice. This examination can require a further round of fact-finding, once again using conventional techniques. Where there is a mismatch between the model and the practice, some improvement may be possible; or, if the model is found to be inappropriate, it may be necessary to return to the systems thinking stage and repeat the modelling exercise.

4.2.1 The SSM Model

The seven main steps to the methodology are normally described in the order shown, and to some extent this sequence is followed in practice; it is worth repeating that it is not necessary to complete each step, nor to undertake them in a set order, and the analyst will of necessity move freely between them or be involved in a number of them at the same time.

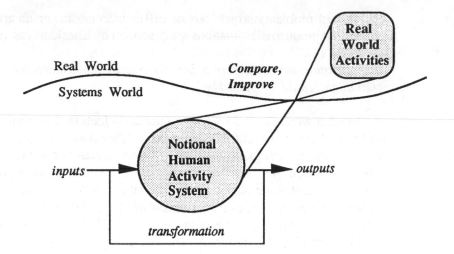

Fig 4.3 - Use of the Systems Model

The flexibility of use, and the idea of continual iteration, are essential freedoms of the approach, and at times the analyst will find it difficult to complete one stage successfully before moving to the next. When this happens, it is better to achieve a comfortable compromise and move on quickly, on the understanding that each and any stage can be repeated when required.

The SSM as a whole is a participative approach, and it is important to involve the client or representatives wherever possible. Operating in isolation during the systems thinking stages can easily lead to the development of ivory-tower ideas and inappropriate models, and involving people from the organisation will not only ensure that their views are reflected, but also help the analysts to keep their feet on the ground. The extent of client involvement will, however, depend on the circumstances of the investigation, and this is discussed at length in Chapter 8.

4.2.2 Problem Situations

The phrase **problem situation** features in each of the first two stages; and it is worth considering what this and the terms **problem solver** and **problem owner** mean in this context. The generally accepted definition of the word 'problem' would be one that implied a solution; a puzzle designed, perhaps, to test the ingenuity of the problem solver, but nonetheless capable of solution once all the pieces have been located and fitted together

Puzzles of this nature are essentially hard problems. For example, a physical link across the English Channel to France is desirable, and it was first necessary to decide what form this link should take, addressing the *what to do* question. Having decided that the link would be a tunnel, the next step is to determine what type of tunnel is most suitable, and how to construct it (and how to obtain enough funds !). These are basically discrete problems, and once the tunnel is completed, the problems are solved. When considering Human Activity Systems, there are many interrelated problems which can be difficult to isolate or clearly define. These problem or problematic situations are described by Checkland as:

"any situation in which there is perceived to be a mismatch between what is, and what might, could, or should be"

In other words, situations where somebody feels that an improvement could be made, without being entirely clear how. Checkland also talks about a *vague feeling of uneasiness* that something is wrong. It could be argued that an analyst or consultant would not be asked to investigate a situation where this was the only concern, at least if the phrase were to be taken literally. However, certain types of studies result from a feeling that improvements can be made without specific ideas at the outset about how these can be achieved; for example, value-for-money exercises, routine departmental reviews, and certain types of computer studies, could all be regarded as situations where there is this *vague feeling of uneasiness* that something could or should be done. It is often felt that new technology can benefit an organisation, a feeling that has frequently led to the 'solution looking for a problem' approach, where favoured equipment and software is imposed on users without a clear understanding of the need for it, the role of the user in the organisation, or how it will aid the enterprise as a whole. The SSM helps to clarify the context of the human activity being examined, and, as a consequence, the overall effect of any technological or other initiatives.

4.2.3 Problem Owners

The term problem owner is defined by Checkland as *"he who has the feeling of unease about a situation"* and also as *"the person or persons taken by the investigator to be those likely to gain most from an achieved improvement in a problem situation".* These definitions are slightly confusing and somewhat contradictory, as the initiator of a study doesn't always gain the most from the results; in practical terms the problem owner can be considered as the person employing the analyst, being responsible for the situation where there seems to be potential for improvement, and who would be instrumental in implementing any change. The problem solver is taken to be the analyst, although it is generally recognised that the owner is at least considering the situation and possibly has ideas that the analyst will take into account.

4.2.4 The 7 Stages of the SSM

Stages 1 and 2 - Expressing an Unstructured Situation (Fig 4.4)

The situation that frequently confronts the investigator when dealing with a soft problem is one where there is this vagueness about what needs to be done; a lack of structure to the problem and the situation that surrounds it. In the general model of the method, this is represented by stage one; more explicitly, this is the stage of a project when the analyst is starting to find out about the study area - talking to the persons involved and other interested parties, reading current and past reports and other documents etc, generally trying to develop a clearer picture of what is going on, and the factors that influence the situation.

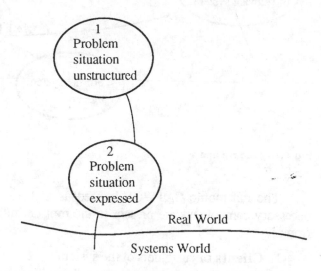

Fig 4.4 - Stages 1 and 2

During this stage a lot of information would be collected about such things as organisation structures, the number of staff employed, processes, locations, names etc, together with the views of individuals and the prevailing issues. As the analyst's knowledge of facts and 'vibrations' increases, the informal structure that binds these factors together will become more apparent. In stage two of the SSM, this situation is expressed pictorially, using a device known as a rich picture, so-called because it reflects some of the *richness* of the circumstances being examined. Taken sequentially, once an acceptable rich picture has been constructed, the investigation temporarily leaves the real world, and the systems thinking phase is entered; however, the picture will be instrumental in highlighting issues that are relevant to the subsequent system models.

Stages 3 and 4 - Defining the Roots/Conceptual Modelling

Before the models are constructed, it is necessary to select a viewpoint, and then define appropriate systems from that perspective. This is achieved by constructing a **root definition** which describes what the system is and what it aims to achieve; not in a strict mechanistic way, but taking account of the persons who could be affected by it, who would be part of it, or who could affect it in some way. It also defines the transformation that could be taking place and the environment that surrounds and influences this particular Human Activity System (Fig 4.5).

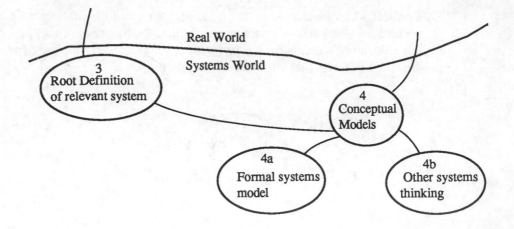

Fig 4.5 - Stages 3 and 4

The mnemonic CATWOE is used as a checklist to help ensure that all the necessary components are present in the root definition. CATWOE is constructed from:

C   **Clients** or customers of the system

A   **Actors** who carry out the activities within the system

T   **Transformation**, ie the change that takes place within or because of the system (ie the conversion of input to output)

W   **Weltanschauung** or **Worldview**, ie how the system is perceived from a particular (explicit) viewpoint - sometimes described as 'assumptions made about the system'

O   **Owner** of the system, ie to whom the system is answerable, and/or who could cause it to cease to exist.

E   **Environment**, ie the world that surrounds and influences the system, but has no control over it.

A series of root definitions and associated conceptual models could be developed, taking a variety of different viewpoints. None of these are necessarily more correct or more appropriate than others, but they should be based on viewpoints that are explicit and can be supported. Each will lead to a different model, which, when used during the comparison stage, will give a new insight into the actual situation.

From the root definition, a model is drawn that shows the minimum necessary activities that must exist for the system to achieve the stated transformation. The model is similar in its intent to the schematic diagram for the jet engine (Chapt 2), ie it is a theoretical or *conceptual* construction, in this case reflecting the analyst's ideas about the defined system, making use of verbs to describe each component or activity.

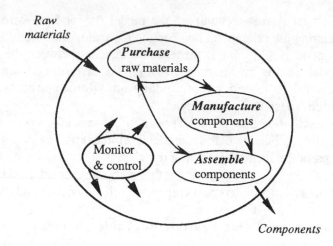

Raw
materials

Components

Fig 4.6 - The Use of Verbs in Conceptual Models

To illustrate the point, a simplistic conceptual model is given in Fig 4.6, showing activities that could be needed to transform raw materials into components. (Although the components of this model could probably be identified in certain organisations, don't forget that at this stage the examination is still taking place in the systems world, and this is a model of what could be, not of what is !)

The model should be checked against the formal systems model described earlier, considering, by discussion, each of the desirable systems characteristics, and how they could be represented in the real situation. Because the SSM uses a fairly high level set of guidelines to structure an investigation, it doesn't preclude the use of any other system models that may be relevant as a basis for comparison. For example, the Tavistock Institute uses a concept of a *task system* which regards human activity as simultaneous technological and sociological systems. If the analyst is familiar with other system models, then they can also be used to check the validity of the conceptual model. (This is shown as sub-stage 4b in the SSM diagram.)

Stage 5 - Making the Comparison (Fig 4.7)

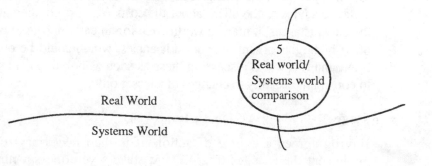

Real World

Systems World

5
Real world/
Systems world
comparison

Fig 4.7 - Stage 5

The next step is to compare the model (or models) with what exists in reality. Crossing the conceptual line back to the real world, the model is used as a form of template to determine if there are mismatches between the actual situation and the model, asking questions such as *do the activities shown in the model exist in reality?*, followed by secondary questions about system characteristics, effectiveness, and efficiency.

Activities that do not appear in the model may also be found in practice, possibly indicating that the model is an inappropriate one. Bearing in mind that the steps of the method are not undertaken in a set sequence, returning to stages three and four might be necessary at this point to reconsider both the root definition and associated model, a process repeated until the analyst and the client are satisfied.

Stage 6 - Deciding Feasible/Desirable Changes

The comparison stage should reveal areas where improvements may be possible ; the next step (stage six) is to examine each of these to determine whether the changes are culturally *feasible* in light of the views, background and experience of the people within the organisation, and *desirable* in systems terms to support the preferred transformation, eg the inclusion of new activities, or strengthening existing ones (Fig 4.8).

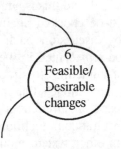

Fig 4.8 - Stage 6

In practice, it is also worth having some idea about the technical and economic feasibility of possible changes, so that the systems dialogue can be reinforced by concrete advice to the client about the costs etc of each alternative. Depending on the circumstances, it may be worthwhile for an early debate to take place; the SSM often highlights organisational deficiencies, which could be outside the particular brief, and it is worth discussing these as soon as possible in case the client wishes to constrain the study to changes of a lesser order.

Stage 7 - Taking Action to Improve

Having agreed a course of action, it is then necessary to consider how to implement the changes (Fig 4.9). At stage seven, the structural and procedural changes are considered, together with changes in attitudes, and more pragmatic matters such as obtaining finance, effect on staff levels, training and so on. It sometimes helps to develop a root definition and conceptual model of a system to implement change, albeit a temporary one, possibly as a form of checklist to make sure nothing is overlooked.

Fig 4.9 - Stage 7

4.3 Outputs and Rules

The SSM, being essentially a set of guidelines for examining a situation, rather than a technique with prescribed procedures, has very few rules in the dogmatic sense of the word. The abilities, skills, and experience of individuals vary considerably, likewise the situations being addressed, and there can never be adequate guidelines or rules to cover all eventualities, nor should there be if the professions of problem-solving are to remain respected, and attractive to those who enjoy challenge. It would therefore be invalid, and unwise, to propose a set of procedural rules that govern the approach. However J Naughton in *The Checkland Methodology: A Readers Guide* (Open University Press, 1977) suggests certain **constitutive** and **strategic** rules that provide further guidance for putting the ideas into practice. The **constitutive** rules, described by Naughton as those which *"must be obeyed if one is said to be carrying out a particular kind of enquiry at all"*, distinguish this approach from others, and can be summarised as:

- The complete methodology is a 7-stage process
- Each stage from 2 to 6 has a defined output, ie:

 Stage 2: rich picture and ideas about relevant systems
 Stage 3: root definitions of relevant systems evaluated by CATWOE criteria
 Stage 4: conceptual models of the systems described in the root definitions, built by assembling and structuring verbs (usually expressed in diagrammatic form)
 Stage 5: agenda of possible changes (derived by comparison of conceptual models with the rich picture expression of a problem situation)
 Stage 6: changes judged with actors in the situation to be (systemically) desirable and (culturally) feasible

- Conceptual models should be checked against root definitions and the formal system model

- Conceptual models should be derived logically from root definitions *and from nothing else*
- Conceptual models are *not* descriptions of systems to be engineered (although stage 6 *may* yield a decision to engineer a system)

The **strategic** rules, described as *"those which help one to select from the basic moves .... those which are 'good' or 'better' or 'best"*, taken to be rules that assist an analyst with planning and applying the SSM, are outlined as:

- Preliminary exploration is conducted by searching for elements of structure and process and examining the relation between the two
- Exploration is *not* conducted as a search for systems in the problem situation
- Exploration may be facilitated by asking 'resource allocation' questions:

    (a) What resources are deployed in what operational process under what ....?
    (b) How is this monitored and controlled ?

- Problem themes, that is, blunt statements of one or two sentences, are used to focus attention on interesting and/or problematic aspects of the problem situation.
- Iterate, especially around the sequence *relevant system-root definition-conceptual model-relevant system*
- Set up stage 5 with important actors in the problem situation

(At this point in the book, the precise meaning of some of these rules may not be clear, but they are worth noting for reference purposes when the more detailed explanations of the SSM are given in following chapters.)

4.4 Summary of the SSM

This chapter has provided a brief introduction to the Soft Systems Methodology, together with some general points that are relevant to its application in real situations. From now on the emphasis will be on *how to do it*, exploring each stage in greater detail and giving more specific advice about the practice of SSM, and examples of where it has been used. Before doing so, it is useful to summarise the main points about the methodology covered earlier in this chapter. These are shown in the table in Fig 4.10.

Summary of The Soft Systems Methodology
It provides a set of *guidelines* for examining an organisation with a view to clarifying where improvements may be possible.
It does not require strict adherence to procedures or rules, although there are certain *constitutive* and *strategic* rules which assist with its application in practice.
The main difference between the SSM and other approaches is the specific inclusion of *systems thinking* stages.
It makes an explicit distinction between *real-world* and *systems-world* activity.
Many of the actions taken by an analyst using SSM are conventional fact-finding activities.
Although illustrated sequentially, can be used in any order the analyst requires.
It encourages a process of *iteration* as the analyst's knowledge increases.
It encourages the analyst to examine the situation from a number of different *viewpoints*.
It establishes the basis for a *debate* with the client about possible changes.
It is essentially a *participative* approach, but can still be of value even if this participation has to be limited.

Fig 4.10 - SSM Summary

5 Expressing an Unstructured Situation

5.1 Introduction

The Collins Standard Reference Dictionary defines the term *unstructured* as:

"not formally or systematically organised; loose, free, open, etc."

which may appear slightly contradictory when applied to situations where, on the face of it, there are many activities and groups that are formally or systematically organised, particularly as the term organisation itself tends to give this implication. However, bearing in mind that we are considering soft situations where the main ingredients are human behaviour, which cannot be predicted with any certainty, and complex personal relationships rather than the formal ones that are reflected on paper, then the definition appears perfectly valid. In addition, it also summarises quite aptly the initial impressions that are formed when entering a strange situation, ie it first appears as a loose collection of activities, people, equipment, files, offices etc, without any clear idea of the underlying form or pattern that binds them together.

This, of course, is not unique to investigative work; everybody encounters strange circumstances from time to time, where it is difficult to get an overall view of the multitude of new factors that present themselves. Stepping off a plane at an unfamiliar holiday destination we are bombarded with a series of images; sounds, smells, people, colours, and so on, each of which makes sense on its own, but is, at the time, seemingly unrelated to the others. After a few unsuccessful attempts to find the local shops, occasionally wandering into the wrong hotel room, or possibly buying a map and deliberately setting out to explore, these alien factors become more familiar, and an overall mental picture starts to form. Almost intuitively we begin to comprehend the pattern that connects all the elements, our understanding of this pattern increasing as we learn or experience more about the situation. Within a few days it is normally possible to start giving directions to strangers, to wander off without getting lost and so on until, by the time of departure, we have a good idea of the locality, environment, people, climate, and the relationships between them. On arriving home, we probably spend the next few months boring our friends, family and neighbours with our impressions of the holiday, impressions that owe more to our intuitive understanding of the relationships than to the individual facts that are reflected in the holiday snaps and souvenirs that are bought back.

In many respects, these reactions are similar to those that occur during the early stages of any investigation, either when facing a totally new situation, or one that is being examined from a new perspective. Interviews are carried out, background papers are read, new activities are observed, and we wander around for a while getting physically and mentally lost. At the same time, however, information is gathered, knowledge is acquired, and impressions start to form. Being a soft situation, it will never become *structured* within the defined meaning of the word, but it is nonetheless possible to find out more about the inherent relationships that are there. This is the function of the first two stages of the SSM, ie 'finding out about' the problem situation that is being addressed, not simply by gathering factual material, but also by expressing the findings in a manner that encourages an overview of the complex relationships that do exist.

5.2 Exploring and Expressing the Situation

The learning process usually starts after some dialogue with the potential client when a view is expressed that something is wrong or that some improvement to a situation is sought. In many cases, the client may not be entirely sure of what needs to be done and may just have a *vague feeling of uneasiness* (referred to in Chapt 4) that something is wrong, or that matters are not as they could or should be. Whatever the brief for the study, the first step is to investigate the situation (Fig 5.1) by using conventional fact-finding methods, such as interviewing, observing, reading up on background material and so on, until the stage is reached where there is a need to express the influencing factors in a cohesive or meaningful way. This expression could be achieved by writing a summary that covers firstly the facts (eg about the processes that are being undertaken and the existing structures, etc), and secondly any observations about problems or issues that are affecting the organisation. It would then be necessary to critically examine these factors and draw conclusions from the text about possible ways forward.

However, any human situation is extremely complex and cluttered, and many pages of manuscript may be needed before an adequate description could be prepared. Even when this has been achieved, the limitations of a textual description, which could be spread over many pages of a report, can make it difficult to understand the relationships between the scattered pieces of this puzzle, or to gain an overall impression of the prevailing atmosphere or climate. Experienced analysts will realise that, by this stage, some of the mist that tends to gather at the beginning of a study will start to lift, and, like the holiday situation, a mental picture forms, and instinctively many of the issues, vibrations, political nuances etc are recognised. It may, however, be difficult to make these explicit from the narrow view presented by normal textual media.

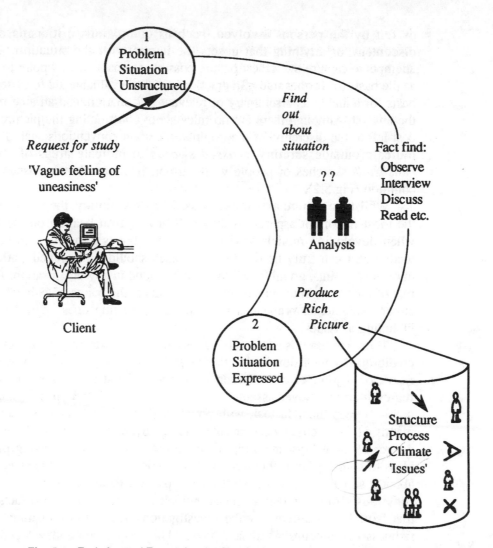

Fig 5.1 - Exploring and Expressing the Situation

This intuitive process can be assisted, and to some extent, expedited, by expressing all the relevant factors pictorially, which groups them together in a concise form and allows them to be viewed as a whole so that the climate is more obvious. This illustration of the 'thousand words' otherwise needed for a written summary should eventually form a picture that reflects some of the richness of the situation, ie containing a wealth of information and indicating areas that are fertile for improvement

A rich picture is a sketch or diagram, usually hand drawn, that depicts certain important aspects of the situation. These might be elements of the **structure**, ie those factors that are slow to change, such as the organisation structure, and physical characteristics such as buildings, locations etc; parts of the **process** that are carried out within the system (constantly changing factors), and how the factors gell together as the **climate**. Also included would be the **issues** that are expressed or felt by the persons involved, such as complaints, criticisms, feelings of discontent, or anything that upsets the harmony of the situation. The picture attempts to capture the richest picture possible to enable a viewpoint to be selected as the basis for further study. In practice it is also a valuable aid to understanding, a basis for a dialogue, a summary of relevant information, and an aide memoire for the analyst. Although there are no rules about constructing the picture, this being best left to the discretion of the originator, certain conventions such as *eyeballs* to indicate outside scrutiny, *crossed swords* to indicate areas of conflict, and *matchstalk* sketches of people with cartoon balloons showing issues, are quite common (Fig 5.2).

Whilst the picture as a whole is useful as a summary, the issues are arguably the most important aspects, as they will have a firm bearing on the later stages when developing models to compare with what is happening in reality. Only factors that can truly be regarded as issues should be included, rather than the minor complaints, grumbles etc that can be found in any organisation. As a general rule of thumb, if, in the judgment of the observer the points of disharmony indicate either fundamental problems, or frequently occurring ones, then they are worth including as issues.

There are hazards to be wary of; assumptions made on too little evidence gain credibility by inclusion in the picture and wherever possible they should be the result of a joint effort between a number of analysts, or with the client or representatives. Never assume that your picture is the *right* one, and wherever possible keep them factual, with opinions shown only where the source can be identified, or when they can be substantiated by sound argument. There is also the danger of the analyst imposing a structure on the picture, allowing preconceived ideas to colour judgement. Whereas it is difficult to remain objective, particularly about a situation that the analyst is now part of, there should be a deliberate and conscious effort to retain a neutral outlook and show only those facts and issues that have been gathered by the investigation. In addition, too much detail and richness can obscure significant points. The early pictures developed during the Case Study described in Chapter 10 were too cluttered to be meaningful, and only showed that the situation was extremely complex, making it necessary to summarise the detail before certain fundamental points became clear.

5.3 Drawing Rich Pictures

The idea of illustrating pertinent facts and issues about a set of circumstances in pictorial form provokes a variety of reactions from the initiated, from comments about childish scribbles and colouring books, to general misconceptions about the pictures being some form of system model. Most persons who have an enquiring mind eventually realise the value of this technique, albeit after some education and persuasion, although some remain unconvinced and prefer to summarise relevant points in a purely textual manner. It is surprising how reluctant newcomers (to the SSM) are to start on these sketches, seemingly inhibited by a blank sheet of paper and possibly concerned in case they 'get it wrong'. The notion that an unscientific device that is not governed by strict rules can be of value appears to conflict with long-held beliefs formed after many years of applying hard techniques to problems, where answers are generally predictable and the approach fairly mechanistic.

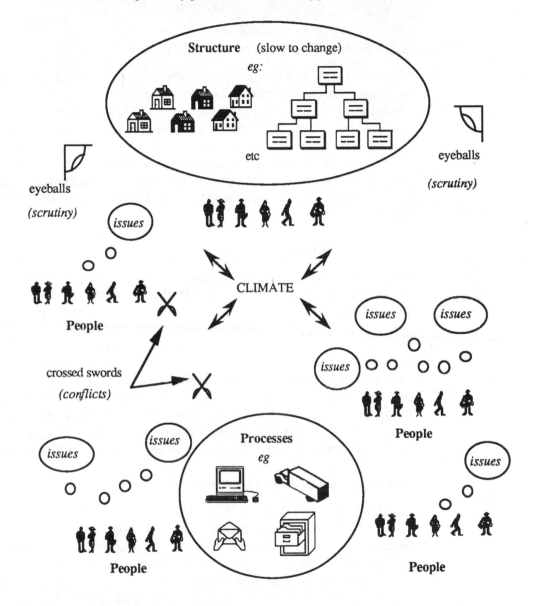

Fig 5.2 - Characteristics of Rich Pictures

The advantages of rich pictures are difficult to put into words (it's a bit like trying to describe the aesthetic values of paintings such as the Mona Lisa!), and there is little to be gained by attempting to do so; the advantages only become obvious once the approach has been used. However, it is indisputable that they are of benefit, and it is always surprising how readily new ideas and views emerge as the pictures are formed. They are not part of an exact science and it is obviously not possible to lay down hard and fast rules about their construction. Lessons can be learned, however, by considering examples from the wide variety of pictures that have been used successfully in studies, and examining in detail the way in which some of these pictures were constructed.

5.3.1 Picture Styles

Checkland provides certain guidelines about the content of a rich picture, suggesting that they should include structure, processes, and issues, and give an overall indication of the prevailing climate. These elements are not explicitly obvious in all pictures, and there are almost as many styles of pictures as there are analysts, each with their own preferences for technique, conventions, and to a large extent, content, depending of course on their particular interest in the situation being examined. Many pictures are drawn intuitively, without a deliberate attempt to formally include the recommended factors, which nonetheless are featured in some way.

A Picture of Vice

One example is shown in Fig 5.3, taken from *Dealing With Complexity* by Flood and Carson, and drawn as part of a study of the problems of vice in central London This picture has some built-in amusement value, but accurately reflects the main issues of policing an area that is notorious for drugs, prostitution, and vagrancy, together with the effects on local residents and tourists. Formal organisation structures are not shown, but the picture does include other *slow-to-change elements*, such as the geographical areas of the West End, shops, hotels etc. The processes of vice are implicit, but the picture is mainly concerned with highlighting the variety of issues that may be relevant when developing system models at a later stage. A key was added to the picture for clarification, which reduces the clutter which can become a problem when attempting to summarise a complex situation.

The Distance Learning Situation

A different approach is shown in the picture of the Distance Learning Unit situation (Fig 5.4) which featured as part of an exercise using the Multiview methodology, described further in Chapter 6. Elements of structure are included, such as the various institutions concerned with the Distant Learning process (South Bank Polytechnic, Manpower Services Commission, and the Paintmakers Association), the Departmental boundaries within the South Bank Polytechnic, and the geographical dispersion of the students. Processes are implied in the wording on the picture, *tutoring* students, *reports* made to the Commission, and *transmitting student records* to the Administration centre. The climate is a reflection of the relationship between the *processes* and the *structure,* and some indication of potential problems can be gleaned where existing structures do not appear suitable for the processes being undertaken.

KEY	A Shepherds Market Area	JPU	Juvenile Protection Unit
	B Sussex Gardens Area		Clubs Office
	C Bayswater Road Area	SOS	Street Offences Squad
	D Piccadilly Circus Area	JAR	Juveniles at Risk
	E Victoria Station Area	WCC	Westminster City Council
	F Soho Area	LTP	London Teenage Project

Fig 5.3 - Rich Picture of Vice in a London Area (courtesy: Flood and Carson, <u>Dealing with Complexity</u>, Plenum 1988)

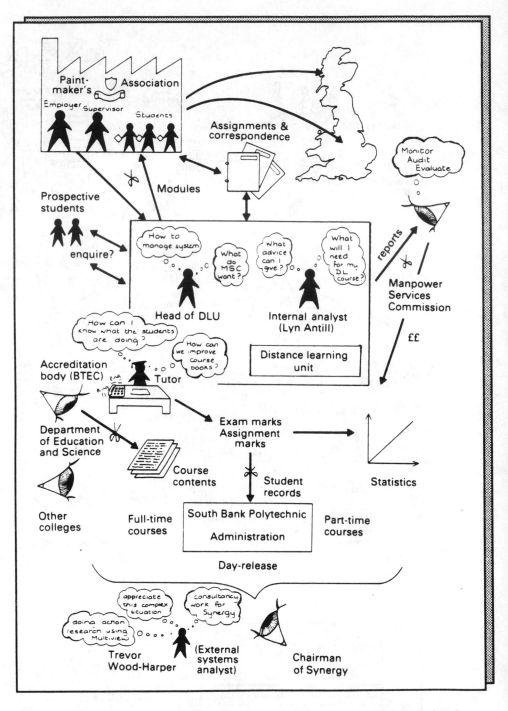

Fig 5.4 - The Distance Learning Situation (courtesy: Wood-Harper et al, <u>Information Systems Definition: The Multiview Approach</u>, Blackwell Scientific Publications 1985)

The issues are illustrated in a fairly conventional manner by the balloon-like structures coming from the heads of main characters included in the picture.

It is worth pointing out that making sense of another person's picture is sometimes difficult, without the opportunity to know what was in the analyst's mind when it was constructed. The main aim is to understand your own pictures, together with those of colleagues working on a joint project, and to be able to explain them to clients when necessary. The second more obvious point is, although a lot of pictures are computerised or typeset before publication, most of them are drawn freehand, and attempting to draw them in the first instance on a computer can be inhibiting, placing artificial constraints on the picture construction.

5.3.2 Putting Pen to Paper

One effective way of making the transition from purely textual descriptions to graphic illustrations is to summarise a section of prose in picture form, possibly taken from a newspaper clipping, and then return to it after a few hours and try to recall the information contained in the original text by examination of the picture alone. For example, try drawing a picture that illustrates the main points of the following, noting that there are no correct pictures; the criteria for deciding their value being the information they contain, and whether or not this information can be recalled by the illustrator.

" *Mr David Mellor, the Home Office minister, announced after meeting President Zia here last night that Britain is to give a further £2.4 million to help stop the flow of heroin from Pakistan, the country which potentially has enough heroin to supply the world as well as the 80 per cent of "brown sugar" narcotic now reaching the UK.*

The contribution brings to £3.4 million in British assistance to opium poppy eradication schemes in Pakistan. It will be administered through the UN fund for drug abuse control and is devoted to a crop replacement programme in the Dir Valley of the North West Frontier where the illegal cultivation of poppies continues to thrive as much as it did when the plant was banned from farms."

The content of any picture will depend on the viewpoint of the analyst, and what information is considered relevant to the investigation. The initial pictures are generally constructed from information gained during discussions when setting up a study, or by reading background material, together with early thoughts about issues that may be relevant. Once something has been committed to paper, the rest follows fairly easily. Noting down the main points from various related papers in diagrammatic form is a useful way to start, for example, by sketching the organisation structure, together with the disposition or location of the groups involved, if this seems to have a bearing on the investigation.

Consider the development of the rich pictures used in the Case Study discussed in Chapter 10. A full description of the organisation is included in the chapter, but in brief, it is a Social Services Department providing support to dependent people in the local community, eg the elderly, mentally or physically handicapped persons, children and families in need of counselling.

The department was controlled by a Director, responsible for implementing the policies of a Committee of elected County Council Members. He was supported by a Management Team of Assistant Directors, and the policies implemented through an *operations* organisation comprising a headquarters section, and 14 Area Offices, each covering a separate geographical area of the County. At the outset of the study, little else was known about the situation, but there was sufficient information to start drawing a picture. (The originals for the final picture shown in Chapter 10 were produced as hand-drawn sketches in the usual manner, but, for purposes of clarity, sections of the picture have been recreated here using computer graphics.)

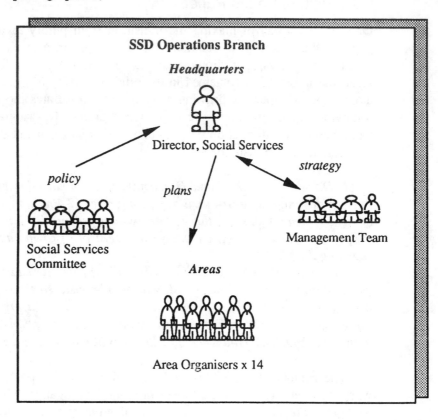

Fig 5.5 - The Start of a Picture

After searching through committee papers, and considering the points made during the early interviews, further relevant information was added, such as the overlap of the Operations Branch with others in the same department, and other Council functions that had an input to Social Services. In addition, it was felt necessary to show that the organisation was subject to the constraints imposed by central government legislation, and would always be under the scrutiny of the public and the media, shown as an overlooking *eyeball*. As a rule, the analysts are also included, as a reminder that the situation is invariably affected in some way by their involvement.

Some indication of processes was also shown in the early pictures, particularly those related to new technology which was the underlying reason for the study. As the study progressed, administrative staff at the Area Offices were included, together with front-line operators (Fig 5.6).

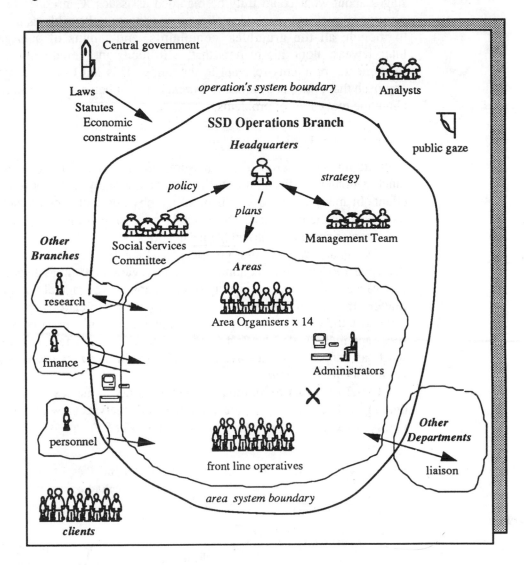

Fig 5.6 - Adding the Detail

In this particular picture, after long and heated debate it was decided to include provisional system boundaries, due to the difficulties in distinguishing between the roles of the headquarters and area offices, and other organisations providing similar caring services, but outside the control of the Council. This tends to *bend the rules* about keeping the real-world situation and the systems thinking exercises completely separate, but in the circumstances of the Case Study it was considered justified because of the complexity of the situation.

5.3.3 Deciding the Issues

In many studies, the issues are readily identified, particularly if the organisation is relatively small. In this study, however, there was some confusion in the early stages about what could truly be regarded as issues. Consequently, the first rich pictures became extremely complicated and incomprehensible as the analysts tried to include all the grumbles, complaints, and moans of a large number of interviewees, alongside matters that, with hindsight, could truly be regarded as fundamental problems relevant to the study. This led to the development of an approach that I have called *interview analysis*, that helps when making a judgement about the root cause of problems.

Interview Analysis

There is no easy answer to the question of what is or isn't an issue, being very much a matter of discretion on the part of the analyst, in consultation with the client. In many cases they will be fairly obvious, but when examining a large organisation, the process can be aided by reviewing points made by interviewees, and summarising them into problem categories.

This is achieved by first grouping together, from interview notes, key points that appear to have some commonality, then reviewing them to determine a possible root cause. For example, a variety of comments were made about relatively minor matters that could be traced back to poor communications in the department. Similarly, there were a lot of comments made about such matters as 'missing client files', and individuals who deal with the same client at different times not knowing what had happened previously, all of which could be summarised as a *breakdown in the flow of client data*.

Initial attempts to summarise minor points as issues were undertaken by drawing a series of mini-pictures, reflecting the analysts view of the root cause of a number of key points, such as those shown in Fig 5.7.

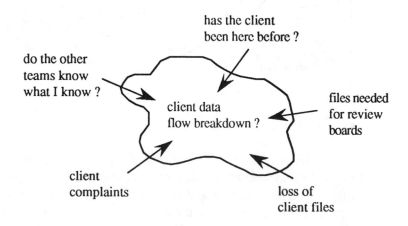

Fig 5.7 - Picture Summary

After a while, this exercise became quite laborious and time-consuming, and, taking account of the fact that the initial interviews had produced something in the order of 400 key points, extremely inaccurate, depending almost entirely on the memory of the interviewer, and a large volume of interview notes. As a result, a flat-file computer database was utilised to summarise each key point against a predetermined code indicating the possible problem area.

This technique was developed from an earlier Organisation and Methods study of the needs of a number of departments for office service support, eg typing, photocopying, filing, and so on. Representative staff were asked a series of questions about these services, and each reply given a code to indicate the service being discussed (ie T - typing, F - filing, etc). The key points were then manually extracted from the interview notes, and grouped within the coded categories so that an overall summary of views about the services could be prepared (Fig 5.8).

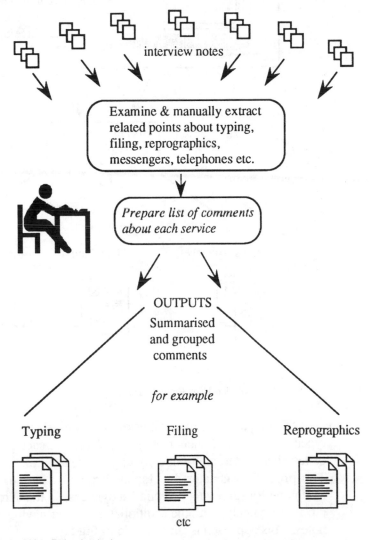

interview notes

Examine & manually extract related points about typing, filing, reprographics, messengers, telephones etc.

Prepare list of comments about each service

OUTPUTS
Summarised and grouped comments

for example

Typing   Filing   Reprographics

etc

Fig 5.8 - Manual Key Point Analysis

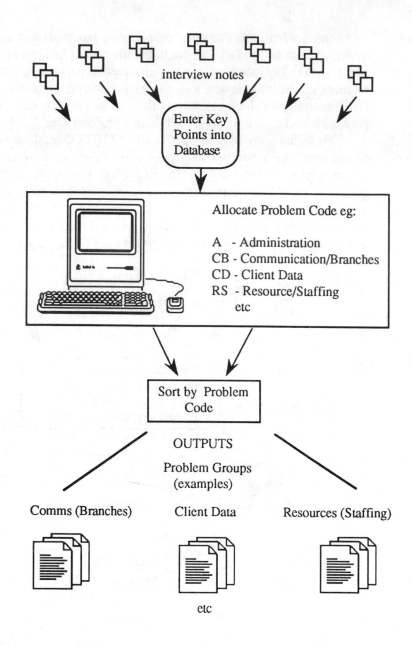

interview notes

Enter Key
Points into
Database

Allocate Problem Code eg:

A - Administration
CB - Communication/Branches
CD - Client Data
RS - Resource/Staffing
etc

Sort by Problem
Code

OUTPUTS

Problem Groups
(examples)

Comms (Branches)   Client Data   Resources (Staffing)

etc

Fig 5.9 - Computerised Key Point Analysis

After a period of trial and error, it was found that the interview notes could be typed as a series of statements into a database, each point given one of the above codes, then sorted alphabetically to produce a summary under each coded heading. The process was later developed further as part of a three person interview technique for surveying potential study areas, so that a report for the client could be produced quickly, with the additional benefit of having a tidy copy of the interview notes. This approach is summarised in Fig 5.9.

Figs 5.10a and 5.10b are extracted from the database used to record and sort the interview notes raised during the Case Study, giving an example first of the notes entered in interviewee order with a predetermined code added. The second example shows the result of an alphabetical sort on the coded field, grouping together points on similar matters.

Author	Key Point	Code
CL	Client files get lost in the system	CD
CL	Policy implementation & evaluation reports - time-consuming	PE
CL	Time delay for referrals to reach teams	CO-S
CL	Poor support from general office	A
CL	Insufficient clerical back-up	A

Fig 5.10a - Key Points in Interviewee Order

Author	Key Point	Code
CL	Client files get lost in the system	CD
MD	Need for client cross-reference system	CD
DH	Mentally ill cases taken on already known to home helps	CD
JC	Senior staff also need access to client data	CD
JE	Card-index checks not carried out	CD

Fig 5.10b - Key Points after Sorting by Code Field

These summarised key points were then incorporated into the rich picture as areas of potential issues. Unlike the office services study, where it was easy to identify related points about the various services, a great deal of judgment was required when deciding the relationships between comments on less tangible matters. Nonetheless, the technique provides a form of mechanical assistance for a process that is essentially judicial regardless of the methods that are used.

Despite this summarising process, the final picture was still extremely complex and covered a full A3 size page. For the sake of clarity, and for ease of reproduction, certain factors about the organisation that were either less relevant to the study, or more obvious to the analysts and client, were removed. This had the effect of 'de-cluttering' the picture, which was then redrawn on a microcomputer, in the process highlighting those facts and issues that appeared to be most significant. However, this composite picture would not have been a true representation without the originals having been hand-drawn in the first instance, bearing in mind that the technology will inevitably influence the way in which the picture is constructed.

Fig 5.11 reflects some of the main points identified by the analysts, with the full picture reproduced in Chapter 10.0

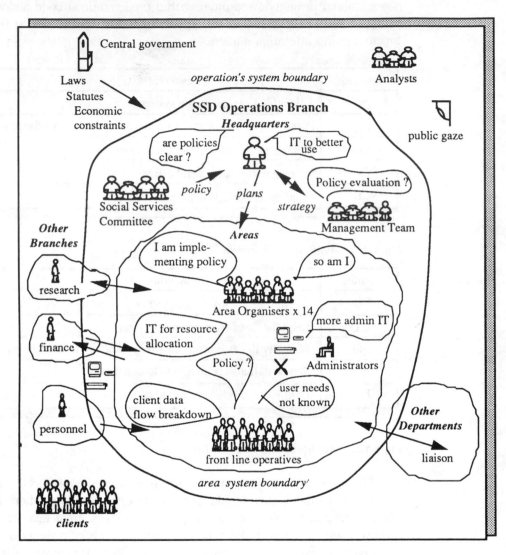

Fig 5.11 - Extract from Case Study Rich Picture

Lessons Learnt

There are certain lessons to be learnt from this particular exercise. Firstly, if the organisation is large, covering a wide range of multi-functional activities, then a certain amount of focussing is required in the early stages. Ideally, a broad, global, view of the situation should be taken initially, identifying, with the agreement of the client, issues that are worth examining in more detail. The accessibility of the client is obviously a major factor at this stage, a point made in earlier paragraphs.

In the Case Study, the status of the client and his 'busy diary', together with a view that it was the analysts job to report back once the issues had been identified, resulted in the team members progressing the study without regular contact.

Secondly, there tends to be an apparent explosion of problems in the early phases of a major review, particularly if the selection of interviewees results in all main functional groups being represented, each with their own viewpoints, and areas of discontent. In this type of situation, the analyst can be extremely distracted and pursue seemingly relevant points that, taken in the context of the whole, are not fundamental issues. A matter for judgment again, which can be aided by the computer analysis technique.

'Fundamentality' and Frequency

Fundamental is a word to keep in mind when deciding whether or not a point made is truly an issue. Consider whether the matters raised in discussion indicate some fundamental problem at the heart of the organisation, eg poor communication, poor control, unclear policies, and so on. In addition, it is worth looking out for those problems that are in vogue, or of particular concern *at that time*. In the social services study comments were made at all levels about the Home Help service, the **frequency** of which indicated that this should be regarded as an issue, and included in the picture for further consideration. 'Fundamentality' and frequency, although not mutually exclusive, are useful catchwords to help the analyst make a judgment about matters raised during the fact-finding stages.

5.3.4 Using the Pictures

The essential idea of using a picture that reflects the richness of a situation is to provide not only a concise summary of pertinent factors, but also to allow the analyst to select viewpoints from which to develop system models, ie to identify Human Activity Systems that appear relevant to this particular situation. In many cases, as the picture is drawn and the analysts give explicit consideration to the prevailing issues, ideas begin to form about notional models that could be constructed. The statement that the SSM is not carried out in a strict sequence is quickly verified as the study progresses; all the stages form part of series of actions and thought processes that are going on simultaneously, with the rich pictures providing a useful base to move out from, and return to when checking the validity of conclusions about the situation. Examples of the way in which pictures help in this and other respects are given in the chapters describing various applications.

Apart from the formal use of the pictures, they undoubtedly provide an extremely useful aide memoire, which can be conveniently carried on a clipboard for easy reference, or kept close to a telephone so that the analyst can impress unexpected callers with a wide range of knowledge about one or more extremely diverse subjects !

situation @ present (ie constraints)

RP used to identify systems that go into CM. — model (of components) necessary to achieve transformation

described in RD.

5.4 Conclusion

This chapter has covered in some detail the initial stages of a study using the SSM, where the real-world is explored, and then *expressed* using the medium of a rich picture. Checkland states that the function of stages one and two is:

"to display the situation so that a range of possible and, hopefully, relevant choices can be revealed", adding that *"that is the only function of these stages"*.

Which sounds quite clear and simple, although, because there are no rigid or mechanical procedures either for constructing the picture or deciding the relevancy of choices or viewpoints, it is not quite so easy in practice. These matters and others are discussed further in Chapter 6 that follows.

6 Modelling Human Activity Systems

6.1 Introduction

The development of Human Activity System models is the cornerstone of the systems approach, underpinning all the other stages of the investigation. The concepts are not difficult to grasp, and anybody that has experience of using diagrammatic techniques for representing man-made systems, or recording clerical or computer procedures etc, will already have the basic idea of the form of the models, and what they aim to portray. Having said that, it is generally the most difficult stage of the Methodology, and one that analysts are frequently least confident about. I have often been tempted to sit down and invent a board game based on Stages Three and Four of the SSM; something akin to Snakes and Ladders, where the action goes back and forwards and little progress is made, or possibly Scrabble, with endless debates about the meaning or validity of words. The process of constructing root definitions and the subsequent models is without doubt the most intellectually demanding aspect of the SSM, causing much soul-searching and discussion (accompanied by frayed tempers and strained relationships!), but also giving a tremendous insight into real-world problems.

Systems modelling is the activity that most closely resembles a technique, following certain rules about construction and content. However, because what is being illustrated is a concept that may not have a tangible manifestation in practice, and the choice of system components is largely left to the imagination and intuition of the analyst, there is never a perfect model, nor can its existence be fully verified by examination of what is actually going on. The use of plain language to describe the system components leads to problems with semantics and interpretation, and prolonged discussions that, at the time, seem to be distracting and counter-productive.

Nonetheless, it is this part of the process that produces the goods, even with less than perfect models and related doubts about their validity. It is essentially a 'lateral-thinking' exercise which encourages the observer to stand back from a situation and deliberate it in a detached manner, an exercise which inevitably increases the observers knowledge, and stimulates thoughts about improvements that might otherwise have been overlooked.

It is worth remembering throughout these stages that the root definition, and the associated model describe a notional system, ie a system that, although relevant to the situation, is only an idea in the mind of the analyst. It isn't an attempt to describe what actually is, this being one of the functions of the rich picture, but a description of what might exist from certain explicit viewpoints.

6.2 Root Definitions

Root definitions were introduced earlier on, with a brief description of how they are constructed and what they are used for, ie to describe the system that is about to be modelled. At this point it may not be clear why it is necessary to list all the CATWOE elements, when models can be constructed on the basis of the transformation alone, an approach that is explored in later chapters. Put simply, it encourages a better understanding of the situation that is being examined; in the process of formulating the root definition, matters are deliberated that might otherwise be overlooked, resulting in the analysis being too shallow.

For example, are the *actors* in a situation just those people employed by the organisation being studied? They may be the only ones who are directly affected by changes that result from the analysis, but there could be others whose work overlaps that of the organisation, and who are part of the system being considered. An education system may well include activities carried out by the parents of students; a community care system could include social workers, voluntary agencies, the District Health Authority, relatives, friends and the police etc - the study may only be directed at and affect one enterprise, but the involvement and influence of others should be recognised if the analysis is to be a thorough one. At some point, it may be necessary to disregard actors etc who are outside the control of the client, but this decision will be taken in full awareness of their existence. Careful consideration of the CATWOE elements ensures that all these relevant factors are made explicit, putting the client organisation in the right perspective.

For ease of reference, the process of developing a root definition is examined prior to the discussion about constructing conceptual models. However, it should be borne in mind that the root definition is not an end in itself, but the basis for a subsequent model, an idea that can be summarised as the 'root from which the conceptual model grows' (Fig 6.1).

6.2.1 Primary Task and Issue-Based Root Definitions

There are two fundamentally different types of systems that can be described by a root definition; ie **primary task** and **issue-based**. The first is that considered necessary for an organisation to fulfil its most obvious role, eg to manufacture (specific) goods, to provide education, to sell newspapers, and so on. In effect, the definition reflects a neutral account of this role, and most of the related activities should be present in practice, otherwise the organisation could not survive. It is normally possible to identify the outputs from primary task systems, and from sub-systems that have contributory roles; *cars* would be produced by a car-manufacturing system, and related sub-systems would produce outputs such as raw materials, plans, assembled components, painted panels, and so on. The organisation structure may not reflect the model, but the activities and the people responsible for them should be clearly recognisable.

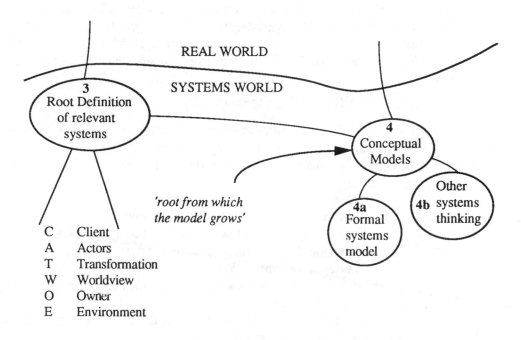

Fig 6.1 - Link Between Root Definitions and Conceptual Models

Issue-based systems reflect a particular point of view, rather than a general or public one, and the existence of such systems could be considered contentious. For instance, although a primary task model could be constructed that reflects the role of management services in a company, Checkland suggests that :

"not being a wealth creator, (management services) will continuously have to justify its existence by demonstrating that the cost of providing its efforts is less than the benefits that flow from them"

implying that, from certain points of view, the existence of a management services system will always be an issue. Similarly, a prison could be regarded in contentious terms, eg to provide punishment, to protect society, to reform criminals, to maintain a criminal sub-culture; each description being open to question and potentially an issue. At the same time, a prison could be described in primary task terms, possibly to *receive, store and despatch criminals.* All Human Activity Systems can be viewed in a variety of ways; the primary task view tends to be more detached and devoid of any sociological aspects.

Deciding whether or not a root definition is issue-based or from what viewpoint it could be regarded as contentious is often difficult, and tends to be a bit academic; the only clear distinction is between between primary task definitions and those that aren't ! Even when deliberately setting out to construct a primary task model, invariably a number of provisional statements will be devised that it is difficult to agree on, and in this sense, primary task systems are inevitably preceded by issue-based analysis. Experience has shown that it is seldom possible to prepare a definition that has universal agreement, hence my earlier advice, that to avoid too much distraction, be willing to accept a comfortable compromise on the understanding that the models can always be adjusted as the study progresses.

6.2.2 The CATWOE Components

The CATWOE mnemonic, as discussed earlier in Chapter 4, is one suggested by Peter Checkland and used by many exponents of the SSM as an aide memoire to the factors that make up a well-formed root definition, ie:

C **Clients** or customers of the system

A **Actors** who carry out the activities within the system

T **Transformation**, ie the change that takes place within or because of the system (ie the conversion of input to output)

W **Weltanschauung** or **Worldview**, ie how the system is perceived from a particular (explicit) viewpoint - sometimes described as 'assumptions made about the system'

O **Owner** of the system, ie to whom the system is answerable, and/or who could cause it to cease to exist.

E **Environment**, ie the world that surrounds and influences the system, but has no control over it.

Although the mnemonic is not intended to suggest a sequence for developing the root definition, it can encourage users to define first the Customers, then the Actors, and so on through the list. In practice the most important elements of the description are the *transformation* and the associated *worldview*, and these need to be determined before other components can be properly identified. A rearrangement of the letters to form TWECOA (or other variations starting with TW) might be more appropriate to provide guidance for inexperienced users of the methodology, but this would break with the established custom and practice amongst soft systems exponents.

The CATWOE mnemonic also tends to infer that each element of the root definition has equal importance; however, defining the transformation together with the associated W is arguably the most crucial step, being the basis for the construction of the conceptual model that follows. Experienced analysts will no doubt empathise with the idea that, having defined the change that the system brings about (ie the *processes* that it undertakes), the related inputs, outputs and sub-systems can also be identified, allowing an embryonic model to be constructed.

6.2.3 Transformations

Early clarification of the transformation is one of the clues to successful completion of the systems thinking stages, being the basis of the activities represented in the subsequent conceptual model. Although the idea of converting inputs to outputs is easy enough to grasp, deciding what they are and what transformation is involved is not so easy when dealing with the abstract notion of a Human Activity System. There are similarities to the exercise of identifying inputs, processes and outputs when developing a computer specification; for example, the inputs of a customer order and a prescribed format would be merged together or *processed* to produce an invoice for payment, ie the output (Fig 6.2).

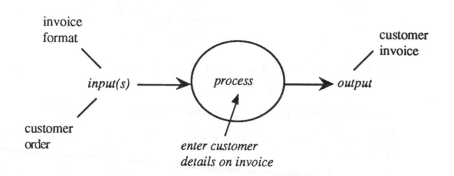

Fig 6.2 - Representation of an Observable Process

These are, of course, observable activities, which is where the similarity ends. With Human Activity Systems, abstract notions are being considered which won't always have tangible form in reality. Nonetheless, when attempting to define the transformation, it often helps to think on the lines of 'input - transformation - output', using the same simple sketch to develop ideas. It is sometimes useful to decide first what the output should be, then determine the necessary inputs, leading to a statement of the change that the system will bring about, ie the transformation. Consider the notion of a *customer satisfying system*; the output being a *satisfied customer* from an input of *customer needs*, the transformation being to *clarify and satisfy customer needs* (Fig 6.3).

Simple examples are useful to illustrate a point, but in practice deciding the transformation can be a challenging task, particularly as there will inevitably be different levels of inputs/outputs, some of which may not be apparent until the modelling activity is further advanced. To assemble the resources to satisfy customer needs, some tangible input may be required, such as raw materials etc, or even information to decide how the needs could be satisfied. These can be regarded as secondary inputs to the system as originally defined, and will produce secondary outputs at the same level of abstraction.

It is important to express the transformation in a balanced way, ie if the input is in conceptual terms, then the output should be the same. In other words, a conversion of an abstract notion to a tangible form is not physically possible, and consequently is not supportable in a root definition.

Fig 6.3 - Representation of Human Activity Transformation

6.2.4 Weltanschauung or Worldview

Most people have difficulties with the concept of W, difficulties that are not easily resolved by studying the various textbook definitions, such as Checkland's *"the (unquestioned) image or model of the world that makes this particular human activity system (with its particular transformation process) a meaningful one to consider"*, or the more succinct *"assumptions made about the system"* suggested in the Multiview publication (Blackwell Scientific Publications, 1985).

To clarify what is meant by these phrases, it is useful to start from the premise that human activity can be considered from a number of different viewpoints - the point from which you view a system, or indeed any tangible object, dictating what you see, or perhaps would like to see. Physical constructions such as buildings can be viewed from above, the side, end, or from an angle, and these perspectives then illustrated by an engineering drawing as side and end elevations, plan and possibly isometric views. This analogy could be extended to include the different perceptions of the same building from the viewpoints of an architect, artist and user, the architect seeing it in terms of geometry and structural design, the artist in terms of its aesthetic value, and the user in terms of functionality (Fig 6.4). Although these are essentially conceptual perspectives, if each person were given the brief to design and construct the building, it is likely that the end-products would be significantly different, achieving the same purpose but with a change in emphasis.

Similarly, it can be argued that individuals will perceive human activity from different viewpoints, depending on their background, experience, and particular interest in the situation. Their perspective (defined in this instance as "a mental picture of the relative importance of things" - Concise Oxford Dictionary) will vary accordingly.

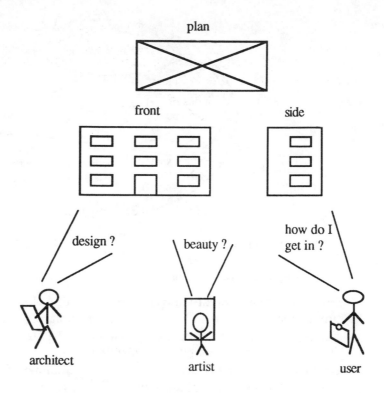

Fig 6.4 - Viewpoints/perspectives

Bearing in mind the earlier discussion of overlapping systems in organisations (Chapter 3), an observer could be concerned with the primary task system, or the information system, or industrial relations and personnel systems, or even the system for arranging the office Christmas party (Fig 6.5)! More contentious views could also be relevant; the enterprise might be regarded as a *system for polluting the environment* (ie with chemical waste or other pollutants), or *for making enormous profits for the shareholders*.

Returning to the definitions, by considering whose or what viewpoint is being taken, and the perspective from this viewpoint, we can formulate ideas about the system that this implies. (Taking an obvious example, if a study is being undertaken by a communications expert, it is most likely that the communications system of the organisation will be of prime concern.) It is then necessary to decide if this is a relevant model for these particular circumstances; for instance, a system for selling newspapers is hardly likely to be found in an undertakers, but is relevant to the *image* we have of a newsagents, or even to a hospital, or a railway station, ie situations where it is reasonable to assume that such a system might exist. By  explicitly considering W, therefore, the analyst is validating the transformation specified in the root definition, ie by stating that this 'image' or these 'assumptions' make it reasonable to suppose that the transformation might be occurring.

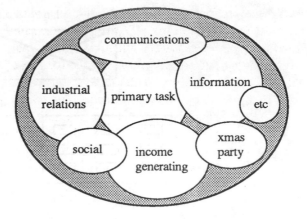

Fig 6.5 - Overlapping Systems

Selecting Relevant Viewpoints

When selecting relevant viewpoints, it is worth making a list of those persons or groups that may have an interest in the situation, based on an examination of the rich picture and other information. Customers, shareholders, the managing director, staff and so on, may well have different perspectives, and the analyst should consider the systems that could be conceived from their points of view. For day-to-day projects, it is also useful to explore the activities that are required to achieve the main purpose of the organisation before more contentious views are considered, ie by first developing a primary task model from a neutral or public viewpoint. This is never as simple as it sounds, complete neutrality can seldom be achieved; even such statements as *to convert raw materials to motor cars* or *to provide residential accommodation for elderly people* could be the source of argument, and issues of one kind or another will inevitably feature in the discussion.

However, modelling a transformation related to the primary purpose of the organisation is often a prerequisite to developing ideas about other overlapping or complementary systems, particularly when considering information or communication aspects, both being by-products of the organisation's main activities. A primary task model also has a bearing on other issue-based models that might be derived; to properly explore the provision of social amenities in a public house requires some consideration of activities such as buying and selling stocks, calculating costs, employing staff, etc.

As far as the work of consultants is concerned, some of the viewpoints could be considered more relevant than others; if employed to improve cost-effectiveness of the operation, advice on the industrial relations system may not be welcome, unless it is a contributory factor to increasing profits or reducing costs. To reinforce this point, consider again the management services function discussed in Chapter 3. In some organisations it could be (and often is) viewed by manual workers as a means of keeping wages to a minimum, but by the Managing Director as providing a pool of intellectual and other resources to solve any problems that management may have. Taking the manual workers view and constructing an appropriate model will be revealing and educational, but may lead to ideas about improvements that conflict with those of the MD. This is not to say that other viewpoints shouldn't be explored, but to make the pragmatic point that, when suggesting changes, it is worth giving due consideration to what the client is most likely to accept. (It could also affect the chances of the analyst surviving in the long-term !) As a rule of thumb, when deciding whose viewpoint could be *most* relevant, consider who would need to approve any changes that may be proposed as a result of a study.

It isn't easy to give definitive advice on these and other matters relating to systems modelling, which is one reason why many practitioners do not readily accept the approach, needing the reassurance of correct answers to the types of question that are posed. It is worth remembering that all these devices are just a means to an end, the end being improved understanding of the situation, which then conditions subsequent investigations and associated findings. Any root definition that is reasonably correct will encourage this to happen, and being too pedantic can inhibit progress to a satisfactory conclusion.

6.2.5 The Environment

The requirement to comprehend all the factors that affect a system warrants the inclusion of the environment in the root definition. In some cases this is relatively easy. Taking the definition:

'the world that surrounds and influences the system but has no control over it'

it can be seen that certain types of systems, eg those that are concerned with making a profit, selling goods, calculating prices etc, will be affected by the prevailing economic conditions such as high interest rates, or availability of raw materials, or the current state of the housing market and so on. These factors influence the system, but have no control over it. Likewise public services are affected by the political environment, which, on a day-to-day basis, has no direct control but can cause a major disturbance if the climate changes. Recent pressure on local government to compete by tender with the private sector for certain services (such as cleaning and catering in schools, and the maintenance of public grounds and gardens) is such a disturbance, and will require major changes within the parent organisations. Similarly, the current concern about the environment (ie the 'green' feeling) will cause many corporations to take account of public views about waste, the ozone layer, and the protection of certain species, etc, and adjust their operations accordingly.

Too much disturbance to established bodies can be severely detrimental, and there is always a danger that they will not be able to adapt, and consequently would cease to exist, at least in any recognisable form. The instigators of major disruptions could also feel the backlash; for example, central government could be displaced in favour of a different political party if the electorate became unhappy about the variety of changes taking place, and desired a return to a more stable state of affairs.

These are all important factors that could have a bearing on any associated studies, and by explicitly considering them, the analyst becomes aware of forces that can affect the way in which the organisation operates, and the changes that it is prepared to accept. At a lower level of detail, the environment is sometimes more arduous to define. What, for example, influences, but doesn't control, the way in which a typing pool or the assembly line in a factory operates ? Is it the overall climate of the organisation, or are the local pressures more relevant, such as the desire of immediate customers for a fast turnaround, or the militancies of certain groups of staff that could inhibit the introduction of improved working practices ?

In practical terms, it always helps to keep the simple definition in mind when attempting to define the environment, ie *influences but doesn't control.* On a more general basis, the previous advice still applies; make time to consider this aspect of the root definition, if necessary accept a comfortable compromise, but don't become embroiled in an endless search for utopian answers !

6.2.6 The Customers

It is also necessary to clarify who benefits from, or is affected by, the system that is being defined. It may seem obvious that the customers of an organisation are those that receive the products or services that are turned out, or, in the case of systems with unpopular outputs such as 'punishment', those who suffer as a result. This may well be the case if the model reflects the role of the organisation, but only in terms of the output of the system as a whole. The clients of the sub-systems of the model are invariably the components with which they interact, receiving inputs such as information, resources and so on, as part of the transformation process (Fig 6.6).

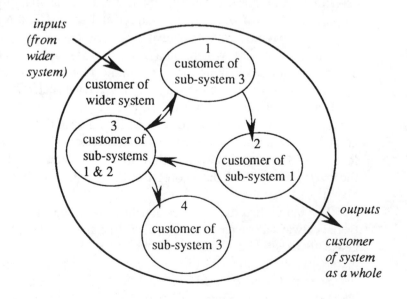

Fig 6.6 - Customers and More Customers

When decomposing the first resolution system model (ie expanding each sub-system, as explained later in this chapter), further root definitions can be formulated for each component, leading to deliberations about these 'internal' customers, and the inputs that are received. The customer defined in the root definition will therefore depend both on the viewpoint taken, and where the system boundary is considered to lie, and it is important to ensure that the customer specified at each level is the one affected by the system defined *at the same level*. This of course raises the question once again of system hierarchies and wider systems of interest, and how to define the limits of the investigation. Despite the best endeavours of experienced analysts, there are no firm guidelines on resolving these dilemmas, but there are some hints which can help the analyst reach a comfortable conclusion, as discussed in the next section on Actors & Owners.

6.2.7 Actors and Owners

In the earlier discussions, the point was made that the actors who carry out the system activities are not necessarily *just* those who are employed by the organisation, but may include people outside of the control of the client. To further complicate the analysis, the actors might also be considered clients or beneficiaries, or even the owners of the system itself.

For example, employees may also be shareholders, or local authority ratepayers; in one respect they are owners of the system, but also receive the benefits that the system gives, or they may be employed in the organisation providing the benefits, and are therefore actors. Apart from such people themselves suffering an identity crisis, analysts have been known to go round in conceptual circles attempting to arrive at satisfactory conclusions when formulating the root definitions. Once again, this is due primarily to the variety of overlapping or nested systems that can be imagined in any situation, with the players changing roles depending on the viewpoint selected. There is also a tendency to mix the reality with the concept; the roles that people take are being debated, not the individuals themselves, as each might wear a different hat for different occasions. Defining ownership gives rise to similar consternation, even with the following definition as a guide:

'to whom the system is answerable, and/or could cause it to cease to exist'

Because of the different levels of system that are nested within the hierarchy, a number of owners could be identified, each with the power to 'shut down' the systems below their position in this hierarchy (Fig 6.7). As these problems arise, it is sometimes necessary to merge the abstract ideas with some down-to-earth consideration of the actual situation being examined.

Using the concept of systems within wider systems, and knowing the way the organisation is actually structured, it is possible to arrive at a compromise that works in practice, based on the premise that the owner of any system is the person (or persons) that needs to be convinced in order to get changes made within their span of control. Ergo, if changes are required to a discrete functional group, the owner could be regarded as the immediate boss; alternatively, if it is felt that an individual in this position would be too involved to take an objective view, then the next person or group in the hierarchy might be a more appropriate owner.

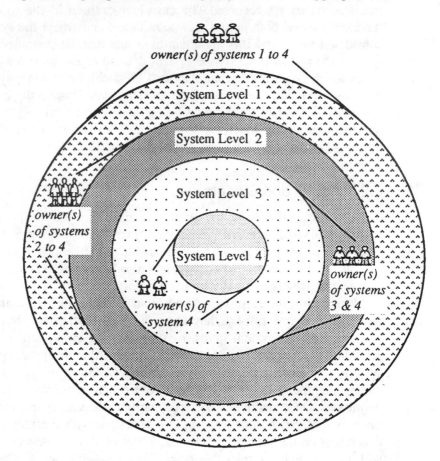

Fig 6.7 - Levels of System Ownership

In the local authority study described in Chapter 10, the project team eventually settled on an ownership hierarchy based on this approach, which could then be extended to cover all possible levels of ownership, from the individual front-line units to the local authority as a whole. In Fig 6.8, the Area Managers who are responsible for the day-to-day work of social workers and support establishments were taken to be the owners of the (care delivery) systems that could be developed at this level. In other words, the Area Managers are assumed to have authority for agreeing and implementing changes that would affect their geographical area only. The department head (the Director) was considered the owner of systems encompassing the branches of the department, including all the Areas, and would need to agree County-wide changes. The elected Councillors *own* the whole service department, and are responsible for fundamental changes to the structure or policy. Conceivably at some point, basic changes in services would come to the attention of the ratepayers (particularly if the changes affected the overall costs!), who could cause the system to cease to exist if not satisfied, and can therefore be considered owners at this level. Taking this approach to extremes, ownership of the whole local authority system rests with central government, who can, and do, affect it nationally.

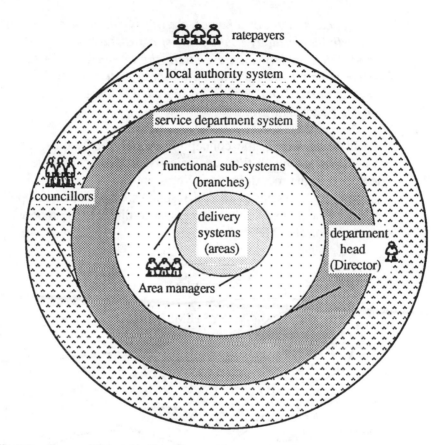

Fig 6.8 - Assumed Ownership Hierarchy

Once again, this approach does not provide the definitive answers, but does provide an acceptable compromise which allows a study to move on, whilst accepting that, should an alternative solution be considered more appropriate at any point, then it can easily be adopted. Identifying suitable actors, owners and clients is desirable to appreciate all the factors that influence system performance, but these factors do not necessarily affect the construction of the model (which is primarily related to the transformation stated in the root definition).

6.3 Conceptual Models ⚹

Having defined the roots of the model, the next stage sequentially is to illustrate it in ⚹ a manner that shows the relationship between system activities. Once again, this precise sequence won't always be adhered to; initial attempts at modelling may well be going on as the root definition is being developed or refined. Various models have been used during the explanation of systems concepts, but it is worth reviewing their construction and content in more detail before moving on to demonstrate how they are derived from a root definition. Checkland defines a conceptual model as:

"a systemic account of a human activity system built on the basis of the system's root definition, usually in the form of a structured set of verbs in the imperative mood. Such models should contain the minimum necessary activities for the system to be the one named in the root definition. Only activities which could be directly carried out should be included - therefore admonishments such as 'succeed' should be avoided"

In other words, the conceptual model shows those components (eg sub-systems or activities) that are logically necessary to achieve the transformation described in the root definition. Verbs are used to identify the sub-systems and/or activities (eg a system to *allocate* resources, or to *implement* policy etc), and there should be no more than those necessary to illustrate the root definition at that level of resolution (Fig 6.9).

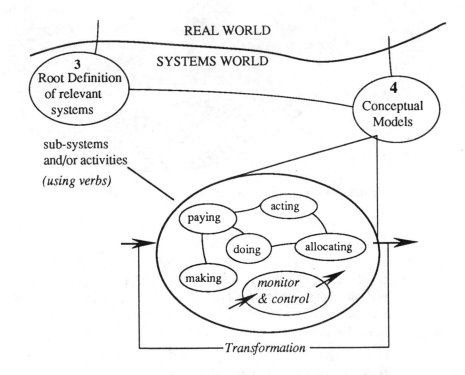

Fig 6.9 - Conceptual Modelling

At the high level stage, the model should not contain too many activities, 5 to 10 is normally sufficient. Any more than 10 probably means that there is a mix of resolution levels; bear in mind that each sub-system can be expanded independently of the first level model. There should not be any component *hanging in the air*; the interactions should always be shown in some manner. As an exception, the monitoring and control element, which relates to all other components, is often included without any specific connections shown, as a device to avoid too much clutter in the illustration.

The activities of the system are usually expressed in terms of **what** is being done, rather than **how**; for example, allocating resources is *what* the sub-system is doing - *how* this is achieved is not shown. *How* may be included in the root definition, but it should be shown as a constraint, such as 'controlling the cost of manual labour *by using Incentive Bonus Schemes*', ie, *how* the cost will be controlled.

6.3.1 Constructing the Model

The point was made in earlier chapters that analysts are often reluctant to put pen (or pencil) to paper and draw a rich picture, particularly when they are newcomers to the strange world of soft systems thinking. The same applies when first attempting to construct a conceptual model; there is often some apprehension that the model may be 'wrong', or at best, misleading. Once again, this reflects the nature of the SSM; because it is not aimed at describing or engineering an *actual* system, but is concerned with encouraging *learning,* the modelling technique does not need to be precise. Consequently there are no 'good' or 'bad' models, although some will be more appropriate and better constructed than others.

It is often said that initial attempts at modelling are undertaken intuitively, a statement that may seem too loose for those who prefer firm guidance on how to proceed on this type of exercise. In practice, however, this is a valid statement, and comes about because analysts are often dealing with areas of human activity that they have some in-built knowledge of, and many models are only a general representation of systems that could exist in the real situation. Experience has shown that this intuitive approach only extends so far, and when the examination reaches a certain level of detail, possibly in the second or third resolution models, it is more difficult, and inaccurate, to 'guess' what components are desirable in systems terms.

At some point an input from persons more familiar with the specialist work of the enterprise is required to ensure that the model accurately reflects what is needed to achieve the lower-level transformations. The implications of progressing from the general to the specific are examined in the following sections, which cover model styles and the process of decomposition to a level where associations can be made with the actual functional groups of an organisation.

6.3.2 Model Styles

An examination of textbooks that refer to the use of SSM indicates that conceptual models have a number of different but acceptable forms. Checkland proposes that the verbs selected are *"structured in sequence according to the logic"*. One method of achieving this is used by Robert Flood and Ewart Carson in *Dealing With Complexity* (Plenum 1988) to construct a customer satisfaction transformation system of a manufacturing company, as shown in Fig 6.10. This type of broad-based model, using terms such as 'compete, employ, produce', etc, could be constructed intuitively, provided the analysts have sufficient knowledge of the situation to which the model relates.

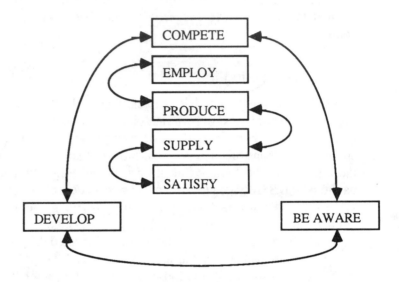

Fig 6.10 - Flood & Carson Style Conceptual Model

Brian Wilson, in *Systems: Concepts, Methodologies and Applications* , uses a slightly different model form, whilst keeping the essential requirement to structure in logical sequence. This is illustrated in the conceptual model of an 'obtain raw materials system' in Fig 6.11, which, although showing lower-order activities, is still fairly generalised.

This style of model is generally most suitable for use where the interactions between elements of the model are too complex to illustrate as a strict sequence of events, and where it is not possible to show the manifestations of these interactions, such as information that passes between each component.These communication links are simply shown as arrows or lines joining the various elements of the model. The MINSE approach discussed in Chapter 11 attempts to define more precisely what these interactions could be, as part of a process of developing information models, and the complexity of these relationships is reflected by the need to use a computer database in support of the MINSE analysis.

A variety of other model styles are illustrated in the relevant chapters of the book, including those used to good effect in a number of case studies, Chapter 10 onwards.

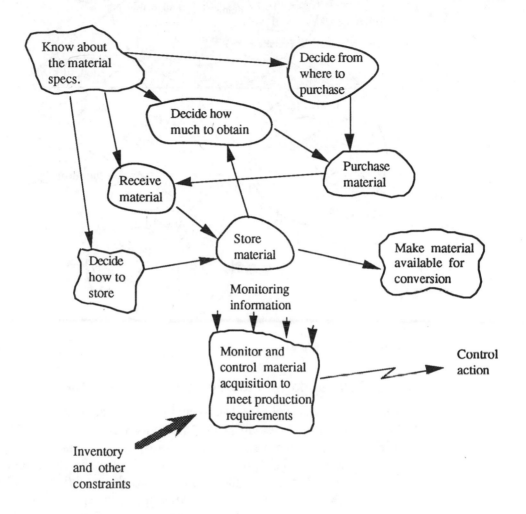

Fig 6.11 - Wilson Style Conceptual Model

6.3.3 Decomposing

Having drawn a first-resolution version of the conceptual model, it may then be necessary to examine further each sub-system to draw out the activities that should be included. The exercise of developing lower-level root definitions and conceptual models (known as decomposing, also carried out in *Structured Systems Analysis*) would then be repeated, bearing in mind that each of these could be regarded as systems in their own right (Fig 6.12).

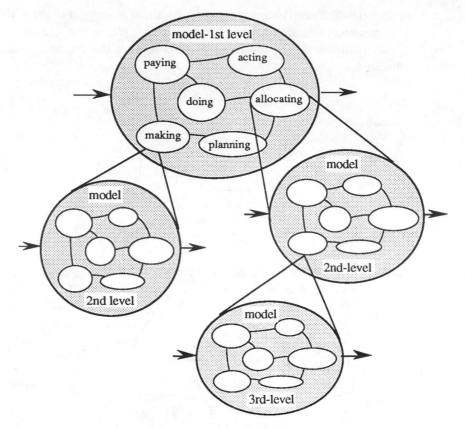

Fig 6.12 - Decomposing

6.3.4 Comparing with the Formal Systems Model

To help ensure that the model is a valid one in systems terms, it is examined in relation to the formal systems model, shown as Stage 4a on the SSM diagram, and covered in detail in Chapter 2. In other words, a model characterising the *general* form of a system is used to check the validity of a *specific* model of a Human Activity System. Effectively, the question *'Does this qualify as a system?'* is being posed, the answer being provided by addressing each aspect of the formal model in turn. As a reminder, this has the following characteristics:

- An ongoing purpose, ie it should achieve a transformation
- A measure, or measures, of performance
- A decision-making or control process
- Components that are themselves systems
- Components that interact
- Existence as part of a wider system or systems, in an environment with which it interacts
- A boundary which encloses the area that the decision-making process has under control

- Resources for its own use
- An expectation of continuity

The transformation and environment have already been clarified during the formulation of the root definition. At this stage the other factors need to be explicitly considered; eg the interrelationships between the sub-systems, which could be based on of the transfer of information, or the movement of materials, etc; how control is achieved, normally by some *decision-making* body such as a committee or management team, or by the managers and supervisors themselves; and so on through the list. Measures of performance will sometimes be difficult to identify, particularly if the system is issue-based, but nonetheless it is possible to derive suitable measures on the basis of the question:

'If this system were to be in existence in the real situation, how could we determine if it is effective'

Depending on the purpose and level of detail required of a particular study, it may be appropriate to prepare a list of the findings for use when examining the actual situation. (During the Procedure Audit process as described in Chapter 13, the resources that the analyst feels are necessary for the system to be effective are listed, and then compared with what is available in practice.) In some cases, when the system needs to acquire resources for its successful operation, this activity would be included in the model; on other occasions the system will make use of given resources, which would be stated as a constraint when formulating the root definition.

Inevitably there will be some overlap between the systems thinking and the real world circumstances at this point. However, this exercise (ie comparing the formal and Human Activity System models) should not be confused with the *comparison stage* of the SSM (ie where an agreed Human Activity System model is compared with what happens in practice), as conceptually they are separate endeavours (Fig 6.13).

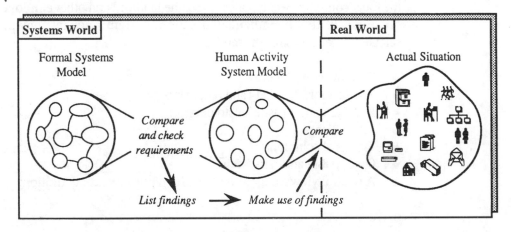

Fig 6.13 - Formal System Model Checks and Real-world Comparison

Bear in mind the approach also recognises the existence of other valid models of human activity (eg the Tavistock model, control and communication models etc), and if required, the characteristics of these should be considered at this point. The systems thinking stages are not in any sense completed once the models are considered satisfactory, as they will inevitably be refined as more ideas occur during the subsequent stages. How the models are used to learn about and improve the real-world is discussed in the next chapter, but, before doing so, it is useful to review stages three and four of the SSM by going through a hypothetical systems exercise.

6.4   The Pub as a System

Using an English-style public house to demonstrate the development of root definitions and conceptual models is common among teachers of soft systems ideas, representing a real-life situation that most people can associate with and understand (with apologies to teetotallers !). It also highlights the point that models of the same situation will vary depending on the viewpoint taken, and, to some extent, the person carrying out the exercise, a fact which is reflected in the different interpretations given in other textbooks using the same example.

In a bona-fide study, of course, the model building stages would be preceded by a fact-finding study, exploring the problem situation and expressing it in rich picture form (Fig 6.14). For the purpose of this exercise, assume that there is an undefined problem with the pub in question, just a *vague feeling of uneasiness* that it could be better, possibly custom has been steadily dropping off, or there is a view that profits could be improved, or even a mixture of both. A comprehensive survey of the circumstances surrounding the pub would clarify the main influencing factors; for example, *slow to change* aspects like the location of the pub relative to its competitors, the relationship between the brewery and the (tenant) landlord, the catchment area for clientele, and so on; *processes* such as delivery routines, 'pulling pints', cleaning up, etc; and *issues* reflecting views of regular and casual clients, the local community and police force, the landlord and other employees.

Even in this hypothetical situation, certain relevant viewpoints could be selected as the basis for the modelling stages, eg:

- The owner of the brewery
- The landlord/landlady and other employees
- Regular and/or casual customers
- The police
- The local community
- Other publicans in the vicinity
- A neutral viewpoint (as the basis for a primary task definition)

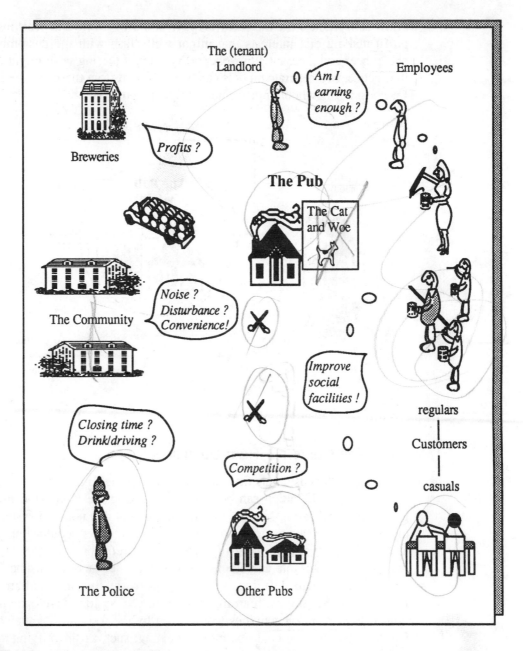

Fig 6.14 - Pub Rich Picture

Each of these could be considered relevant, and would result in a model system that would enlighten the analysis in some way. Taking two of them for example, the customers (both regular and casual) and the employees, including the publican, bar staff and cleaners, we can immediately see that significantly different views of the role of the pub might be held. Customers arguably are mainly interested in the pub as a place providing a pleasant social atmosphere, good service and, hopefully, reasonably priced beer and other refreshments. The employees, however, may be more concerned with the rates of pay, hours of work and working conditions, profit-making etc; interests that might well clash with the customers desire for cheap beer and pleasant atmosphere (Fig 6.15). Ignoring what might be happening in practice, two separate models can be developed from these standpoints as a basis for considering where improvements might be possible.

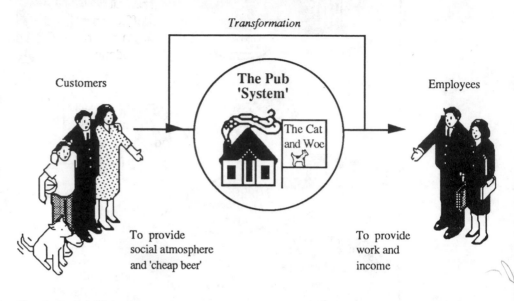

Fig 6.15 - Pub Viewpoints

6.4.1 The Customer Root Definition

Considering the customers first, and in particular the comments reflected in the rich picture, a root definition can be formulated taking the CATWOE mnemonic as a guide, but considering first the transformation that this notional system should aim to achieve. There are a number of possibilities; the system could simply be regarded as one that converts an input of resources (eg refreshments, the building, games facilities, staff etc) into an output of 'social activities' and 'atmosphere', albeit with some difficulty in defining what tangible form this atmosphere might take. However, by starting with an input of resources, system development tends to be limited to *conversion* activities, constrained by the resources available. A broader-based transformation could be more revealing, such as 'identifying and satisfying customer needs (for socialising and a pleasant atmosphere)', probably encompassing a sub-system for obtaining and using appropriate resources to this effect.

With regard to the W taken in this situation, it is possible to see the relationships between the various definitions considered earlier, ie the *viewpoint* taken is that of the *customers*, who *perceive* the pub as a place to enjoy themselves and *socialise,* on the *assumption* that this is an *acceptable role* of a public house, and is therefore a *relevant system* to consider.

Turning to the customers in the root definition, it is valid in this case to regard them as the actual clientele of the pub, being the beneficiaries of the notional system being considered. The actors would obviously include the employees of the pub, and any visiting entertainers involved, but could also encompass the clients themselves, socialising being an interactive affair which would require their participation (active or passive). Ownership probably rests with the publican, being the individual who would have a significant effect on the existence or otherwise of this system, and take day-to-day decisions with regard to its effectiveness. So far the following CATWOE components have been identified:

C - The casual and regular customers of the pub
A - The employees, visiting entertainers, and the customers
T - Customer needs (for socialising) identified and satisfied
W - A pub is a place to socialise and enjoy a pleasant atmosphere
O - The publican

Taking the definition for an environment as 'surrounds and influences, but doesn't control', a number of factors come to mind. The pub must operate within the framework of the law, which obviously influences a notional socialising system, but this is more likely to be taken as a constraint on such things as the hours that the system could be in operation, or the type of socialising allowed (eg no spitting, fighting, dog-racing, or strippers !). The environment in this instance could relate to the current fashion for pub entertainment, such as quizzes, pop videos, pool leagues, and so on, fashion that would influence the system by affecting the customers' opinions about the facilities that are provided.

It is worth repeating that there are no precise answers to the questions prompted by the CATWOE analysis, but it should lead to a root definition that has a 'comfortable' or acceptable feel to it. It sometimes helps to express this as a statement to see if it makes reasonable sense, eg:

'A system owned by the publican, and operated by the employees, visiting entertainers, and customers of the public house, that identifies and satisfies the needs of customers for socialising activities and a pleasant atmosphere, in an environment that influences customers socialising values, constrained by statutory laws affecting the provision of pub facilities'

A conceptual model constructed directly from this definition would be fairly sparse, as it contains only two activity verbs (ie identifying and satisfying), which reflects the high-level nature of the definition. However, it is still meaningful enough to start developing the model, which will encourage ideas about other activities that need to be included, allowing the root definition to be revised if considered necessary.

6.4.2 The Employee Root Definition

Similarly, a CATWOE analysis can be performed for the notional employee system. Consider for this example that the employees' main concern is to 'earn income', and keeping the root definition at the same level of abstraction, the transformation would be 'to identify and satisfy employee needs (for income)'. Considering the W taken, the *viewpoint* is the employees, who consider that the pub is an *income-providing system,* on the *assumption* that it can *provide this income,* and is therefore a *relevant system* to consider.

It is reasonable to consider the owner in this case as the brewery, which could close the pub down, or, by raising the rent, severely affect the ability of the system to provide income at an acceptable level. Once again, the actors are a mix of the paying customers and the employees themselves, each involved in activities that contribute to 'income-providing'. In this case, however, the beneficiaries would be the employees. The system would again be constrained by legal considerations, and the strongest influence could be the competition imposed by other income-sources, such as pubs in the same catchment area, shops and supermarkets, restaurants etc. CATWOE therefore gives:

C - The employees
A - The customers and employees
T - Employee needs (for income) identified and satisfied
W - A pub is an income source
O - The brewery
E - Competition for staff

This could be expressed as:

'A system, owned by the brewery, operated by the customers and employees of the public house, that clarifies and satisfies the income needs of employees in an environment of competition for staff resources, constrained by statutory laws affecting hours of work and tax liabilities'

A further constraint would be required in this definition to recognise that the 'income-providing' activities of a pub are restricted to those concerned with the sale of alcohol, other beverages, and possibly food, and it wouldn't normally be expected to operate outside this sphere, for example, by manufacturing furniture, selling newspapers and so on.

6.4.3 The Customer Model

Taking the root definition for 'satisfying the needs ......etc' of customers, the conceptual model would include the following sub-systems /activities, (as shown in Fig 6.16):

Identify customer needs (for social amenities)
Determine ways of meeting these needs
Obtain appropriate resources
Organise resources to provide amenities
Clarify legal constraints
Comply with legal constraints
Monitor and control the effectiveness of the system in customer satisfaction terms.

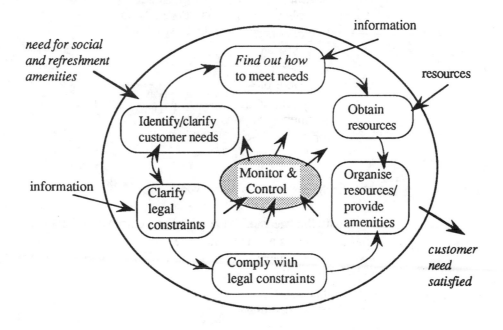

6.16 - Pub System 'Socialising' Model

The next step would be to decompose each sub-system further to clarify what lower-order activities are needed to make the notional system actually work. To identify customer needs, for example, it would be useful to assess the different types of customers the pub caters for, and survey the needs of each client group; to find out how to meet these needs, a visit to other pubs, or a discussion with the Licensed Victualler's Association might help, or even an exploratory chat with a local entertainment agency. By expanding each of the main system components, it is possible to put 'some meat on the bones' and decide what activities should be going on in order for the system to achieve the desired transformation.

6.4.4 The Employee Model

A conceptual model constructed from the second root definition would have some differences; the activities of the first model could be included, but the emphasis changes - customer satisfaction being a means to the end of making a profit.

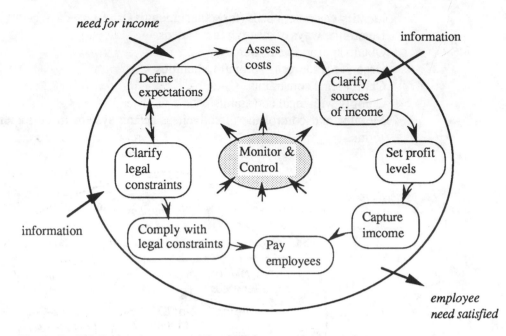

Fig 6.17 - Pub System 'Income-providing' Model

To 'satisfy the needs of the landlord and employees for a source of income' we might have the following components (as shown in Fig 6.17):

Define income expectations
Assess other costs
Identify legal constraints
Clarify sources of income
Set profit levels
Capture income
Pay employees
Comply with legal constraints
Monitor and control the effectiveness of 'income generation'

These in turn would require activities associated with deciding how many staff are required, assessing hours of work, finding out about competitive payment rates; calculating the costs of consumables and overheads, deciding how much customers are prepared to pay, and so on, progressively expanding the model until it is possible to clarify what might be happening in practice.

Before moving on to use these models as templates to explore the real situation, the requirements of the Formal System Model should be reviewed to ensure that nothing is overlooked. In particular, measures of performance need to be derived for the two models to check 'customer and employee satisfaction' respectively, for the models as a whole and for each sub-system. 'Customer satisfaction' may be the aim of the whole system, but, as each of the lower-order activities contributes to this aim, their effectiveness also needs to be measured.

In real life all these components could exist side by side, the main point of the exercise being to encourage learning through the process of taking a cross-section of relevant views of a situation. A root definition describes a system that could exist from a particular standpoint; the necessary activities are illustrated in the conceptual model of this notional system; the analyst then examines the real situation to determine if those activities take place in practice. Those that aren't present, or are noticeably ineffective or inefficient, represent areas of potential improvement.

A pub system could, of course, be seen from a variety of other viewpoints, such as the owners of the breweries, the police, the neighbours, a sociologist, etc, and exploring these would also help in learning about the whole situation. A full conceptual model of the pub could reflect all these overlapping perspectives, and might be necessary if a detailed examination of a real 'pub' was being carried out.

In reality, the two systems examined would be mutually dependent; without customers there would be no income, without income there would be no employees, and therefore no pub system to provide social facilities etc. The difference is frequently one of emphasis, and only relevant to the structure of the model and subsequent examination of the real world. Taking more than one viewpoint is useful for a number of reasons; the analyst's knowledge of the situation is increased by applying some lateral thinking to it; various possibilities are explored to determine where improvements are possible; and subjectivity is avoided when constructing the model(s).

Are both the viewpoints so far considered equally relevant? In this hypothetical case, the answer is probably 'yes'; because the systems are mutually dependent, the pub is unlikely to survive if elements of both were not present in the actual situation to some degree. However, if the analyst is employed by the profit-motivated landlord, then the emphasis would be on those activities that are necessary to maximise income and profit, possibly to the detriment of the customers. Alternatively, a different balance would have to be struck if the review was in the nature of a consumer survey on behalf of customers, or if the pub system was perhaps part of a community centre; customer satisfaction being more relevant than profit. Frequently it is this difference in emphasis that is most important, and may be something that only the client can decide on when debating any changes that the systems exercise indicates are possible.

6.5 Conclusions

Root definitions and conceptual models are intellectual constructs which help an analyst to develop ideas about a situation in a detached manner, and, by considering different viewpoints of a situation, encourage a depth of analysis which would otherwise be difficult to achieve. By making use of plain English as a modelling tool, and being dependent on the interpretation of individual analysts, ideal models seldom emerge, and the exercise of constructing them can cause a lot of debate, frustration, agonising, soul searching etc - all of which can be distracting from the main task of obtaining an end result. Experience has proved that it isn't necessary to be too particular or too pedantic during the modelling stages; all reasonably appropriate models result in an increase in learning which leads to the clarification of problems and ideas about improvements. Checkland advises that:

"On the whole, though, it is better to move fairly quickly to the 'comparison' stage, even if models subsequently have to be refined in a return to conceptualisation".

This is advice with which I wholeheartedly agree. On the understanding that the models can be modified as the study moves on, it is wise to allow set periods of time for systems thinking when applying the SSM; concentrate the mind for that period, then, provided the definitions and models produced are a fair reflection of a relevant system, move on to the next stage and use them to explore what is happening in the *real world*.

7 Exploring the Real Situation

7.1 Introduction

Once the development of conceptual models has reached the stage where they are acceptable to the analyst, it is time to return to the real-world and explore what is happening in practice, making use of the models for comparison purposes. It is important to realise that the purpose of the comparison stage is to uncover potential problem areas for discussion with the client, so that one or more may be selected for further examination. The analyst is not concerned in the first instance with *problem solving* but with *problem identification*, as the basis for a debate about feasible and desirable changes to the client organisation. Finding solutions to the selected problems may well require the use of analytical techniques or methods in addition to the SSM, but these will not be employed until after the initial focussing has been achieved.

It is also worth repeating that the models are not intended to illustrate what is happening in practice, but are just theoretical constructions, and the system activities may not be immediately obvious in the real situation. Furthermore, no matter how well-constructed or seemingly relevant, they do not represent an existing or potential organisation structure, and the sub-systems shown are seldom synonymous with functional groupings. This is partly due to the number of overlapping systems that could be perceived in any situation, with each model representing only one of these; it is also because organisations seldom develop on systems lines, and the models cannot reflect pragmatic concerns, such as the need to group together people with similar skills or for administrative convenience.

So, having spent a long time carefully preparing root definitions and models that seem highly relevant, don't expect to re-emerge into the real world and immediately find evidence that the systems exist in practice. In relation to formal structure charts etc, it is like comparing apples with pears, and it is always necessary to translate what is observed in the real world into systems terms, or *vice versa*, before ideas about improvements can emerge. For example, don't anticipate finding a section of the organisation called *resource allocating* just because that is a component of a conceptual model; allocating resources may be the responsibility of a number of different functional groups, eg personnel, stores, equipment, motor transport etc, and the activities which can be observed might include deploying staff, developing and applying a staffing formula, issuing stores, allocating vehicles to individuals, and so on. Making the connection is often a matter of judgement, which can be facilitated by the methods discussed in this chapter, but which depends mainly on the intellectual skills of the analyst. This is not as difficult as it sounds, and is achieved with a mixture of common sense and experience, with the help of a few guidelines that have been developed from a variety of soft systems studies. Observe what is going on, decide what is being achieved, and see if this relates to the transformation illustrated by the conceptual model.

7.2   The What/How Distinction

One further point needs clarifying; although it is acceptable to describe system components as activities, this can be misleading. Invariably the verbs used during the modelling process indicate *what* is being achieved by the system or sub-system (eg the allocation of resources) and therefore represent a *set* of activities (eg deploying staff, issuing stores etc) which reflect *how* this is achieved. Understanding the implications of the terms **what** and **how** in relation to real-world events is particularly important; the difference between them can cause some confusion, but this has more to do with semantics than with practical interpretation.

An illustration of the difference between the two is shown in Fig 7.1, where the desired achievement (ie the **what**) is *to invest money,* and the methods for doing this (ie the **hows**) can range from putting it in a building society to buying a house. After one of these options has been selected, further *what to do* decisions are required; for instance a house can be bought using cash *or* by seeking a mortgage, which in turn could be obtained through a building society, a bank, or a broker, and so on. This indicates that the term **what** (*to do* or *to achieve*) can be applied at a number of levels; the sub-sets of actions that apply in each case are regarded as **hows**, *at that level.* Translating this into practical terms means that, when models are expanded to show lower-order activities, each of these in turn could represent a more detailed set of actions that could be taking place in the real situation. During the comparison stage, the greater the level of detail achieved during the analysis, the closer it comes to what might be observed going on in practice, a concept that is developed further on in this chapter.

It's interesting to note that, at times, the desired achievement can be identified by working back from the activity level. A recent study into the production of building inspection fire certificates started with a brief to introduce computer-aided design facilities into the associated drawing office. Having completed this, the client was still unhappy about the overall situation, without being sure why. A conceptual model of an ideal system for this service was constructed, and compared with what was happening in practice. This revealed that the customers' views were not normally considered, and, as their expectations had recently changed with the introduction of a fee for a traditionally free service, customer satisfaction was not being achieved. The computer system helped towards this by improving turnround time for producing the certificates (ie addressing one of the **hows**), but only went part of the way to solving the real problem. The modelling exercise highlighted the need to clarify what the client was really expecting in the revised circumstances, and to define an acceptable standard of service. In other words, the desired attainment (ie the **what** in this example) was *customer satisfaction,* rather than simply an improvement in turnround time. Once this was realised, other changes were made to working practices, and proper channels established to take account of the customer point of view, and, in particular, to respond to customer complaints.

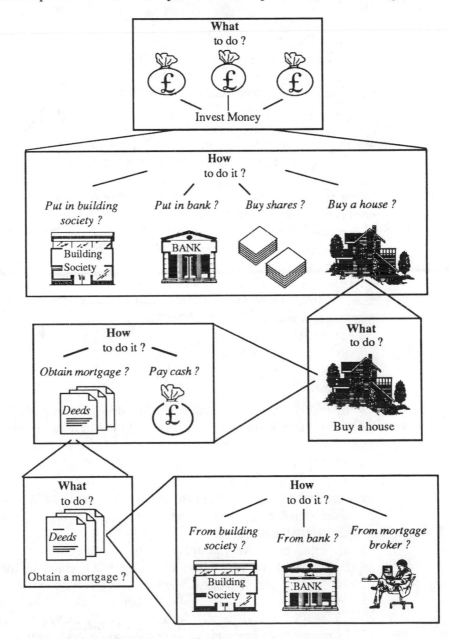

Fig 7.1 - The What/How Distinction

7.3 Making the Comparison

In a theoretical sense, the comparison is carried out by taking the models across the conceptual line shown in the SSM diagram and using them as templates to compare what the analysis suggests is *systemically desirable* with what is happening in practice (Fig 7.2). Where there is a mismatch between the model and the real world, there could be potential for improvement.

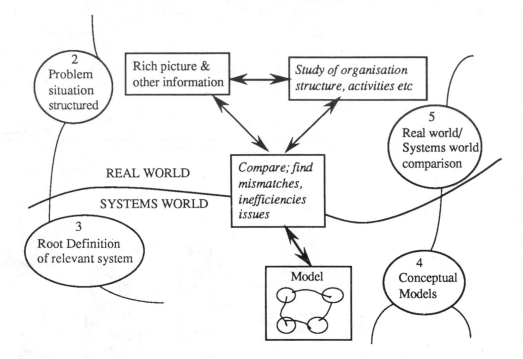

Fig 7.2 - The Comparison Stage

In reality, there is seldom a precise division between the systems thinking stages and the comparison stage; even with the best and most objective will in the world, what is already known about the actual organisation will influence the way the models develop, and, conversely, the process of modelling will itself initiate critical examination of what is already known. Just working through the hypothetical exercise in the last chapter may well have prompted ideas about improvements that could be made to pubs in general, or to particular situations that are familiar to the reader.

In some cases, the models will be instrumental in developing a logical argument about the root cause of issues highlighted in the rich picture. For instance, during the study described in Chapter 10, a great deal of concern was expressed at all levels about policies being unclear. These concerns led to the construction of a model which included a *policy interpreting* component, without which most of the other major activities could not function effectively, and, on further examination, only existed to a limited extent in practice. This also provided a good example of the continual interplay between the various stages of the SSM, with few, if any, ever carried out as discrete stand-alone exercises.

In effect, therefore, the first step towards making the comparison and developing ideas about potential problems can occur almost intuitively during the systems modelling stages. This tends to contradict the notion that systems thinking and real-world activities should not be mixed, but it would be naive and counter-productive to insist that this should never happen; the interplay often provides benefits in terms of expediting the study and satisfying the client.

Once the systems thinking stages are passed, there are a number of other devices that can be used to encourage ideas about changes, ranging from general discussion about the situation, to extensive analysis that explores the links between the systems models and functional groups. These devices or methods are explained briefly in the following paragraphs, and covered in more detail in the chapters covering practical applications.

7.3.1 General Discussion and Observation

It may be obvious that what is going on in the organisation is not what is implied by the model, and this may be drawn out simply by discussion between the analysts, or with representatives from the client organisation (Fig 7.3).

Fig 7.3 - General Discussion and Observation

The feeling that things are not what they could or should be might result from earlier observations, or general experience of organisations of that type, or even from the analyst's own knowledge (eg acquired over a number of years as an internal consultant) about a particular department. Issues illustrated in the rich pictures might be indicators of fundamental problems, and subsequent comparison of the conceptual model with the existing activities could reveal the reasons why these issues arise.

Whilst accepting that models and structures are seldom synonymous, a high-level comparison of the organisation structure with the organisation *implied* by the model can often be carried out to good effect. For instance, if the model suggests there should be a marketing function within the organisation, and this isn't recognised in the structure, it could indicate a possible weakness. The initial high-level analysis could reveal sufficient potential problems to warrant an early debate with the client, at least to prevent the analyst taking a wrong direction or to clarify the client's views on matters that may be contentious. The SSM tends to reveal interesting differences between the systems model and the organisation itself, but pursuing these differences can be very time-consuming and it is necessary to be sure that this is worthwhile and within the original brief.

7.3.2 Question Generation

Another method of making a comparison is by *question generation*, using the models as a source of explicit questions to be written down and answered. It is worth remembering that we are unlikely to be able to see the systems that we consider might exist beneath the surface reality, but are looking for activities that indicate that they are present. Using the pub example, *identifying customers needs* might be achieved informally by talking to the customers about preferences, by listening to and acting on customers complaints, and so on; decomposition of the model could result in related activities being fairly well defined, but invariably the analyst will have to judge whether or not an observed activity contributes to the system as described in the root definition.

It is useful at this stage to draw up a matrix as a form of checklist to focus the thinking on each component of the model in turn. The matrix (Fig 7.4) poses questions about the existence of the activity and how its performance could be measured, as a basis for answering questions about effectiveness and efficiency.

System/ Activity	Exist or not	Measure of Performance	How Done	Assessment	Comment
Identify customer needs	Yes	Level of awareness	Customer survey	Poor	Too casual
Provide Amenities	Yes	Customer satisfaction	Occasional darts matches	Poor	Low-level of amenities

Fig 7.4 - Matrix for Question Generation

Taking the income-generating pub model as an example, we can consider measures of performance in relation to the income produced, eg is it sufficient to meet the needs of the employees and/or the brewery etc, and use these as a basis for determining both the effectiveness of the system, and the efficiency of contributory sub-systems. In this type of situation, it is also possible to develop ideas about who takes the necessary control action should any unacceptable deviation occur (presumably in the pub example only if the income is reduced !). In other instances, particularly where a sociological viewpoint is taken, measures of performance have to be developed on a more subjective basis. The effectiveness of the socialising model, for example, could be measured in terms of customer satisfaction, atmosphere, drunkenness/sobriety, and so on, or even using relative factors such as *more customers each night than the pub next door!*

It is unlikely that any solutions to problems would emerge at this stage; the matrix helps to focus attention systematically on various aspects of reality, formulating further questions for debate or indicating where detailed examination might be fruitful.

7.3.3 Historical Reconstruction

Historical reconstruction is undertaken by comparing what actually happened in a particular situation with an ideal model constructed to achieve the same purpose (Fig 7.5).

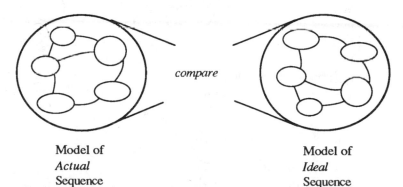

Model of	Model of
Actual	*Ideal*
Sequence	Sequence

Fig 7.5 - Historical Reconstruction

Consider the sequence of events that are required to agree and build a major extension to a large office block. At the conclusion of a project of this size, queries might be raised about costs exceeding budget, facilities not being entirely adequate for the proposed occupants, why the project took longer than originally anticipated, etc. With the benefit of hindsight, a model could be developed illustrating a systematic approach to the problem of agreeing and constructing the extension, which could then be compared to a similar model built around events as they actually took place. This could reveal failings in the approach taken, and lead to a beneficial debate about how to avoid them in future projects.

However, it is worth noting Checkland's comments about this method of comparison, viz:

"This is in logic a satisfactory way of exhibiting the meaning of the models, and maybe the inadequacies of the actual procedures, but it is a method to be used delicately because it can easily be interpreted by participants as offensive recrimination concerning their past performance."

7.3.4 Model Overlay

In many respects, *model overlay* is a way of formalising the general observation and discussion approach, and making it more explicit. In addition to the conceptual model illustrating the root definition for the notional system, a second model is drawn using the same form as the first, but based on the functional sub-systems and activities that exist in the real situation.

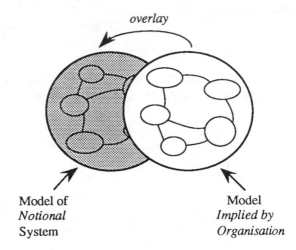

Fig 7.6 - Model Overlay

Overlay of one model on the other (Fig 7.6) reveals areas of mismatch that can then be the subject of debate about change. It also encourages the analyst to consider the root definition that is implied by the model of what exists, and could, inter alia, lead to a re-think of the original root definition and associated model.

7.3.5 Extended Analysis

The methods discussed so far are essentially concerned with making a fairly high-level appraisal of the differences between the model and the actual situation. In many cases this will be all that is required, particularly during the first round of a soft systems study, ie before specific problems have been discussed with the client and one or more selected for further examination. High-level models are simple representations of very complex arrangements of activities, resources, information and communication links etc. In order to explore and understand these arrangements, it is possible to expand such models to the point where all the activities that are desirable *in systems terms* can be listed, and, by focussing on each one in turn, develop detailed ideas about information and resource needs, measures of performance, mechanisms for feedback and control, and so on. This *top-down* examination of the systems model (ie like taking a watch or any other man-made system apart to see what makes it tick), whilst logically sound, needs to be approached with caution; experience has revealed that it can be extremely tortuous and should not be embarked upon unless the client understands and accepts the amount of time and effort involved. However, given these reservations, the outputs from an extended analysis are closely aligned to what can be observed in practice; and at the same time a comprehensive list of the factors that are relevant to the effective operation of a particular enterprise is produced, a list that will remain applicable until a fundamental change occurs.

The theory of extended analysis is not difficult to grasp, once the process of modelling human activity systems is understood. Briefly, each sub-system of the first-level model is decomposed by developing a root definition for the sub-system, and then constructing a second-level model. The process is then repeated for each component of the new model, and so on until the point is reached where the lowest level of activity has been identified, and factors such as information and resource needs can be addressed (Fig 7.7).

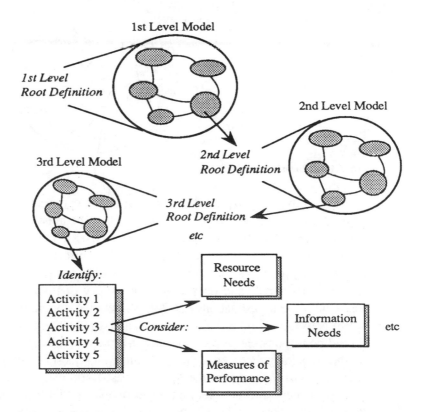

Fig 7.7 - Extended Analysis

Two practical points are worth noting here. Firstly, to avoid losing control of the process, it is necessary to establish some form of *audit trail* so that links with the first resolution model can be maintained. This can be accomplished by annotating the system components with a numerical code which is then extended on a decimal basis to all sub-systems, activities etc (Fig 7.8).

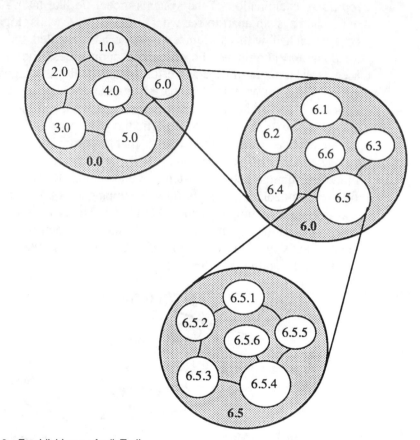

Fig 7.8 - Establishing an Audit Trail

Secondly, the by-products of the analysis can be entered as records into a computer database, which not only ensures that the mass of data that is produced is easily retrievable, but also provides the basis for rearranging the order of the records if required, eg to regroup activities, information items, resources etc. Functional groupings that are ideal in systems terms can be contrived, lists of resources prepared, and questions on a variety of matters can be formulated to assist with the investigation (Fig 7.9). A detailed comparison of system and real-world activities can also be carried out, and a judgement made as to who could or should be responsible for each activity in practice. This latter exercise is a variation on the question generation technique, ie the dual questions of *do the activities exist* and who *could be carrying them out* are posed, followed by further enquiries about how well they are done. (An example of where this approach has been used to good effect is described in Chapter 11.)

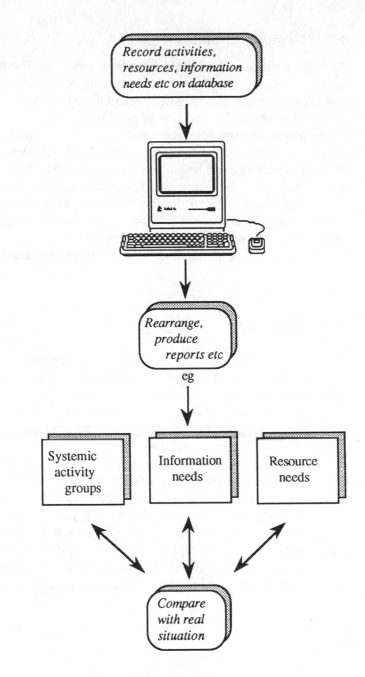

Fig 7.9 - Computer Support of Extended Analysis

Analysis to this level of detail is not always necessary, but when undertaken, helps to solve the main dilemma that faces users of the methodology, that is, how to bridge the gap between the systems thinking and the real situation.

7.3.6 Formal Model Checks

Each conceptual model should have been examined during the systems thinking stages to ensure that it meets the requirements of the formal systems model. During this exercise, explicit ideas will have been developed about how interconnectivity could be achieved, the hierarchy of systems represented by the model, measures of performance, control, resources, and so on. It was stressed earlier that this is a separate endeavour from the exploration of the real world; however, the findings from this endeavour can be utilised during the comparison stage to determine if the *systemically-desirable* factors are identifiable in practice. In particular, organisations are frequently lacking in defined measures of performance, which results in decision-making and control being carried out in an unstructured manner. An understanding of the manner of achieving control etc in a related conceptual model will provide a basis for deciding how this should or could be accomplished in the real situation.

7.4 Summary

The comparison methods described in this chapter are ones that have been derived from a wide variety of soft systems projects, and, as such, provide some valuable guidelines for approaching this phase of the SSM. However, all circumstances are different, and it is often necessary to amend the methods to suit particular studies, or to devise new ways of attaining the same ends; this freedom of approach is reflected in the chapters that follow on practical applications.

In any event, the object of the exercise is to use the ideas developed during the systems thinking stages as a basis for reviewing what is happening in practice, and, after a while, this starts to become second nature and is frequently undertaken without using any formal or prescribed procedures. The output from the comparison stage (ie a summary of the findings and conclusions about what *could* be wrong with the real-world activities) should form an *agenda for a debate about change,* which is then taken up with the client in the next two stages so that suitable options for improvement can be selected and pursued.

8 Agreeing and Implementing Changes

8.1 Introduction

During the comparison stage the analyst has carried out a critical examination of the organisation to determine how it compares with models constructed during the systems thinking exercises. Questions have been posed about the effectiveness and efficiency of the real-world activities, and whether there is evidence of system characteristics, such as good communication channels, measures of performance, and processes for decision-making and control. The relevance of the models to the particular situation should also have been questioned, and the necessary modifications made. From all this questioning and examination, ideas about possible weaknesses and/or potential improvements should have emerged, and the point is finally reached where decisions are required about the direction to take to progress the study, not necessarily to a conclusion, but at least to focus it more precisely.

All studies are different, and in particular, the level of examination will vary considerably, but experience has shown that the SSM can uncover a multitude of problems, some symptomatic of fundamental faults in the client organisation, others of a less significant nature, but still important to the individuals concerned. With all these areas of potential improvement to select from, it is not unusual to feel overwhelmed by this time, accompanied by a desire to return to more scientific methods of analysis where solutions can be selected on the basis of tangible criteria, eg *the costs,* or *the savings,* or *the increase in productivity,* and so on. The main criterion for selecting a course of action when applying the SSM is the client's opinion, based on what is considered to be *feasible* and *desirable* for the organisation in question; seeking this opinion is the role of the final stages (in sequential terms) of the methodology.

There are some further matters to bear in mind when considering these phases. Firstly, the analyst is not attempting to do the impossible and convert a problem situation into a situation without problems; what is required and attainable is to improve it in some way, to make minor or major changes depending on the purpose of the assignment and the wishes of the client. The approach is concerned with *soft situations,* where the problems are ill-defined at the outset, and the first step towards improvement is to decide *what* needs to be addressed. For example, the comparison stage may have indicated that there is a paucity of management information, or that policies are not clear, or customers are dissatisfied etc, each representing an area where the situation could be improved to a greater or lesser extent. The second point follows closely from this, ie the discussions at this stage are not necessarily addressing solutions to the problems, or concerned with *how* to achieve the required improvements (eg by more frequent dialogue between heads of department, the installation of computerised management information systems, and so on), which will be considered once the analysis has reached the level where the problems are clearly defined.

Stages six and seven, therefore, are about **debates** and the action that follows. The comparison has revealed potential areas of improvement, and these form an *agenda for the debate* so that the client can consider what changes are feasible and desirable, in the process of selecting one or more for further study (Fig 8.1).

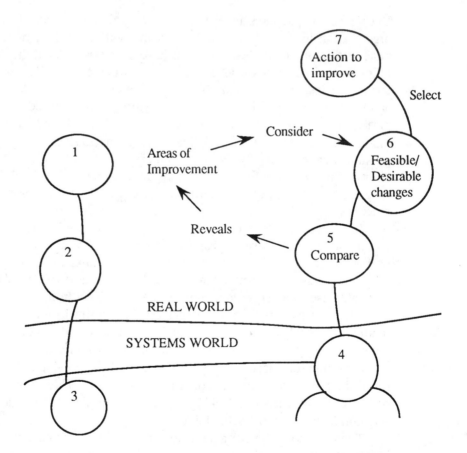

Fig 8.1 - The Final Stages

8.2 Feasible/Desirable Changes

Formally, Checkland suggests that changes of three kinds are possible, ie changes to the structure, in procedures, and in attitudes. **Structure** is defined in this context as those factors that do not change on a short-term basis, such as the organisation and reporting structure, or the make-up of functional groups. **Procedures** are the processes involved in reporting and informing, the activities of producing, and those related to achieving the aims of the organisation. The less explicit term **attitudes** covers changes of influence and the expectations of individuals. In practice, attempting to categorise the types of changes that can be achieved is somewhat academic; the analyst will inevitably be concerned with a variety of specific alterations that could be beneficial, in many respects no different than those considered in the course of other reviews. Exceptionally, because the SSM identifies the problems rather than the solutions, decisions about changes depend more on the outcome of the debate with the client than consideration of factors such as cost-effectiveness etc, although in reality these should not be overlooked, and should feature as part of the discussions.

The term *debate* implies that a conference or forum needs to be held, and where this is practicable, it will undoubtedly be of value. In practice, it may not be easy, or necessary, for a formal meeting of this nature to take place, and factors such as the availability of the client, the geographical dispersion of the organisation, target dates etc obviously need to be taken into account. The initial debate may well be a local one (ie between analysts and client representatives) from which a range of acceptable possibilities may emerge, which can then be discussed further at an informal meeting with the client.

Checkland suggests that the changes that are considered should meet two main criteria. They must be systemically or logically desirable, ie as reflected in the appropriate system model or models, and they must be culturally feasible, given the *"characteristics of the situation, the people in it, their shared experiences and their prejudices."* It may be difficult to agree and implement changes that meet both criteria, particularly the latter, but the exercise of considering them should lead to a better awareness of possible human reactions, and could assist with deciding the best tactics for proceeding. Pragmatically, it is also worth taking account of other factors, such as the *economic* feasibility of proposed changes, given the current financial situation of the organisation, or whether such changes are *technically* feasible, particularly when considering automating existing processes. This, of course, is normal practice when developing proposals or recommendations as a result of any study. It is also worth noting that, because the SSM encourages a *holistic* view to be taken, and many of the influencing factors will have already been considered during the analysis, the analyst is well placed to give advice about the overall effect of certain courses of action, and whether they would be acceptable to the organisation as a whole.

Chapter 10 provides a good example of a study where a full and intensive debate took place with a cross-section of people from the client organisation, covering all the problems that the analysis had revealed. Like most real-life situations, this did not follow any prescribed pattern, and the terms *culturally feasible* and *systemically desirable* did not feature as part of the discussion. However, in this instance and many other cases, the acceptability of certain changes can be assessed from the reaction of the individuals involved, without formally discussing them in relation to these criteria. In most studies, it is unnecessary to hold a formal meeting of this nature, ideas are often discussed as they arise, and the process of selecting areas for improvement is carried out on an evolutionary or progressive basis.

8.2.1 Human Needs Analysis

For routine studywork, it may not be appropriate or practicable for an analyst to undertake an analysis of the personal needs of individuals or groups as a guide to the attitudes that could affect the successful implementation of change. In any event, the soft systems approach requires the participation of people from the organisation being studied; their views and those of others affected in some way by the organisation will have been considered during the early stages of a project and reflected in the rich picture, and, in some cases, in the conceptual models that have been constructed. Experienced analysts will take note of any points that are raised which reflect personal opinions and feelings, and, wherever possible, take such views into account when developing ideas for improvement. It goes without saying that organisations cannot function without people, and it is an unwise manager or analyst that does not take some account of the needs of staff at all levels, particularly when major changes are being considered. These needs are often formally represented, eg by managers and trade unions, and it is normal practice to clarify their views as a prerequisite to the formulation of any implementation plan.

However, circumstances do arise where a more precise assessment of individual or social needs is desirable, and, although it is not the role of this book to explore the many theories of human behaviour that have been developed over the years, it is worth taking a pragmatic look at how this could be accomplished within the context of a soft systems review. Analysts of most persuasions are familiar with the extensive research carried out over the years by people such as Maslow, Mayo, Hertzberg, and more recently by Rudy Hirscheim; details of the literary contributions of these and other researchers is given in the bibliography. A refreshing approach to the problem of assessing human needs, which draws on this research and also gives some clear guidelines on practical applications, is included in the FAOR methodology (see Chapters 9 and 14). This makes use of a *Needs Analysis Instrument,* which, although developed mainly for use when designing technology-based office systems, can be applied to most areas of human activity. It is of particular interest within the context of this book, not only because of its connection with the SSM, but also because of its expressed intent, ie it:

"..aims to be simple enough to be applied by analysts without much knowledge of formal psychological and social analysis methods"

The Needs Analysis Instrument provides a structure for determining first the views of individuals in relation to a variety of factors, then aggregating those views so that the joint needs of groups of people with similar roles can be assessed. It is designed to complement the soft systems approach by exploring the issues relating to social or human factors that are revealed in the early stages of a study, and recognises that human needs are affected by such factors as the working climate, the environment, the work content of tasks, and associated constraints, pressures and responsibilities (Fig 8.2).

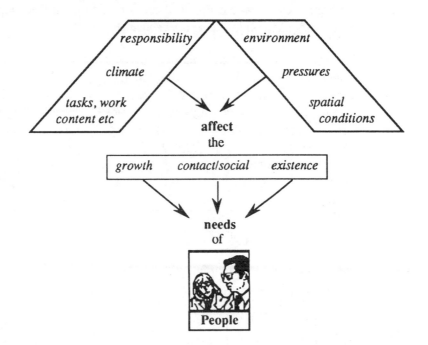

Fig 8.2 - The Rationale of Needs Analysis

In broad terms it concentrates on three categories of needs, ie *growth needs, contact or social needs, and existence needs,* which are explored on an individual and group basis by formulating questions under the following broad headings:

- Task and position
- Job content
- Work performance
- Coordination and cooperation
- Organisation and social development
- Problem-orientated questions
- Questions concerning the person

Each of these forms a module of a comprehensive questionnaire which is completed by individuals who represent groups of people with similar roles. A preliminary analysis of the questionnaire responses is carried out, and a group discussion held to consider the conclusions that could be drawn. This discussion helps to ensure that the questionnaire results are interpreted correctly, and that the reasons for any inconsistencies in the responses are understood. The final phase brings together the results from the pre-analysis and group discussions so that more explicit conclusions can be drawn from the examination.

The approach, although based on the expressed views of *individuals*, is primarily concerned with assessing the needs of *groups* of people, and this process is facilitated by incorporating devices such as the **Likert** scale in the questionnaires, ie where opinions are recorded on a five-step scale ranging from *strong disagreement* to *strong agreement* (Fig 8.3).

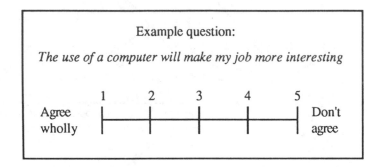

Fig 8.3 - The Likert Scale

This makes it possible to calculate the mean value of all responses to specific questions, indicating the average range of opinions of a group: similarly profiles can be developed reflecting the *divergence* of opinions. Other techniques are used which may be of interest to analysts who are involved in questionnaire design generally, such as the **Kunin-face-scale** (Fig 8.4), used to elicit answers about satisfaction levels, and making the questionnaire easier for the respondent to complete.

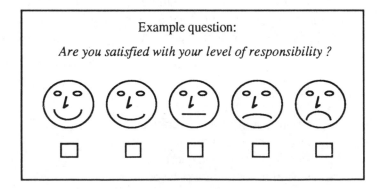

Fig 8.4 - The Kunin-face-scale

The Needs Analysis Instrument is highly recommended for use when an attitude survey is required during a soft systems review. The use of this Instrument is discussed further in Chapter 14, together with an example of the type of results produced and how they can affect the outcome of a study; additionally, a comprehensive explanation of Needs Analysis, which includes a detailed description of the questionnaire techniques, is given in the FAOR publication (ie *Functional Analysis of Office Requirements - A Multiperspective Approach*, by G Schafer - John Wiley and Sons 1988).

8.3 Action to Improve

The discussions about potential improvements will eventually result in agreement to pursue a particular course of action, illustrated as the *action to improve* stage of the SSM. This could be expressed in high-level terms, and it may then be necessary to repeat the earlier SSM stages and focus on a particular problem area before developing appropriate solutions. It is worth noting that a rigorous application of the SSM to a large organisation can be extremely successful at uncovering fundamental weaknesses, and, ergo, areas where major improvements are possible. These may not fall within the scope of the original brief, but it is easy to become carried away by grandiose ideas, spurred on by the intellectual challenge of constructing Human Activity System models and so on, to the extent that this brief is forgotten or effectively ignored by the analysts. As a result, a number of indisputable and well-substantiated conclusions may be reached which do not, however, address the original target of concern. Despite the emphasis on client participation, what happens on a day-to-day basis is the responsibility of the analyst or project leader, and it is necessary to avoid wasting resources by remaining aware of the clients general requirements in terms of results, and tailoring the approach accordingly.

There is also the danger that the examination will concentrate on high-level analysis to the detriment of practical considerations. There is evidence of this *ivory-tower* approach in a number of research studies, where, despite the prime aim being to test out a new approach or a hypotheses, the client's expectations about the outcome have been raised. These are often false hopes, which should either be avoided from the outset, or recognised when proposals are developed, for example, by including possible costs and/or disadvantages of pursuing each option.

To illustrate this point, a recent study which made use of the SSM as part of a methodological package resulted in proposals for improving information storage and retrieval methods in the client organisation, including the introduction of suitable computer support. These proposals were a combination of *what* needs to be done to make an improvement, combined with detailed explanations of *how* this could be achieved. However, little consideration was given to the possible costs involved, and the amount of work needed before firm specifications for the revised systems could be drawn up.

Without this information to balance the apparent benefits of the proposed changes, they were accepted at face value. Subsequently, a full review was commissioned to advance the proposals, a review which eventually resulted in the client only implementing part of them because of the high costs involved. Whilst accepting that some of the responsibility in this instance must lay with the client, who understood the research nature of the project, it reinforces the need for analysts using the SSM to be aware that it can produce unrealistic or naive results, to the detriment of the client and probably the analyst's reputation.

The analyst should endeavour to have regular dialogue with the client as the study progresses. The practicalities of client participation throughout the whole project are discussed later in this chapter, but this can obviously reduce the likelihood of misunderstandings occurring. The soft systems approach is not an exact science, and on its own will not address solutions and associated factors such as costs, but before deciding which changes to pursue it is essential that the advice that is given takes account of all relevant factors.

Once the systems examination of the agreed area of improvement is completed, solutions must be developed for the chosen problems, using other investigative and problem-solving techniques as required. The methods used at this stage will, of course, depend on the type of problems being addressed, and may involve conventional systems analysis and development techniques, or those associated with Organisation and Methods studies. In many respects, the process from now on will be similar to the Operational Research activities covered in Chapter 3, ie:

a. Decide what needs to be done, stating clearly the aim or objective.
b. Determine alternative ways of achieving the objective.
c. Appraise the costs of each alternative.
d. Build a model (if required) of the different alternatives, and test each model under different conditions.
e. Decide, on the basis of pre-defined criteria, the preferred or optimal alternative.

It may also be helpful to construct a notional system model to help with planning the changes, such as that adapted from Brian Wilson's model given in *Systems: Concepts, Methodologies and Applications*. The model, shown in Fig 8.5, has been constructed from the root definition:

"An organisation owned and managed system for cost-effective implementation of a set of agreed changes to existing structures and processes in a way which is acceptable to the organisation and the appropriate unions and which minimises the unintended disruption of the social situation"

Wilson makes the point that the model is only given as a guide and not as a definitive statement of an implementation system. The model is showing an ordered sequence of events for achieving a defined purpose, in this case for *implementing change*, and demonstrates the value of using a systems approach as an aid to methodical thinking, an idea that is explored in Chapter 9.

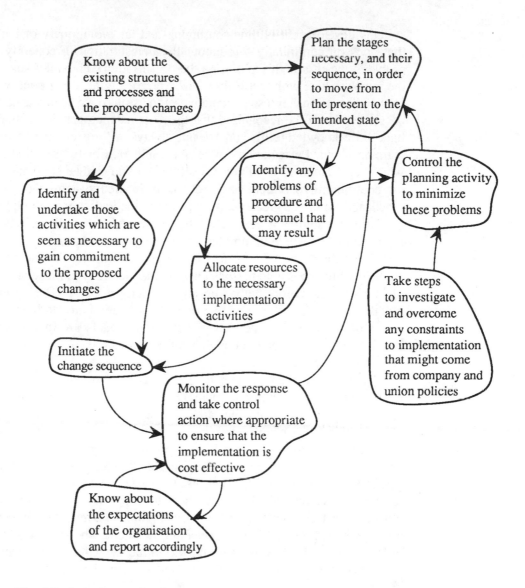

Fig 8.5 - An Implementation System

8.4 The Analyst/Client Relationship

There is a great deal of emphasis on the involvement of the client in a systems review, from the opening stages, throughout the development of models, and when reviewing feasible and desirable changes, and so on. This should help to ensure that suitable viewpoints are taken when developing root definitions, and the validity of models, and make it easier to decide which actions to pursue when considering changes. Undeniably in an ideal situation there would be client/analyst dialogue continually, or at least a representative of the client working in close partnership with the analyst. In many cases this ideal cannot be realised, particularly as all systems development is time-consuming and an evolutionary or iterative process that goes on continually throughout the investigation. Frequently the client or problem-owner is highly placed in the organisation's hierarchy, and, having made the initial contact with the analyst whilst preparing the study brief, will not expect further discussions until some tentative conclusions have been reached.

It also has to be recognised that the systems approach is one that most clients have little knowledge of, and are not always willing or able to devote time to learning about. Presentations of results using rich pictures and systems models, which can be tailored to suit the level of client knowledge, are quite successful, but the continual involvement of the client in detailed systems work isn't normally practicable without a greater understanding of the underlying principles.

In many studies a general briefing is arranged for all persons who may be involved in the review, to introduce the participants and provide the opportunity for explaining the proposed conduct of the study. Following this, detailed discussions are held with responsible persons lower in the hierarchy to make an initial selection of interviewees, before embarking on the study in detail. During the interviews, points about the structure and role of the organisation are clarified, and eventually issues are uncovered. The rich pictures and models are then developed in the background, ie between the members of the project team, and later discussed with subsequent interviewees.

During such discussions there is a tendency among analysts to be reluctant to display their rather untidy scribbles, and a feeling that these are alright for use amongst the project team, but anyone else would find them either silly or too esoteric. What is needed of course is some time spent on educating the client and the staff about the approach being taken, but once again this is not always practicable. Despite misgivings about displaying these 'works of art', reactions are generally favourable, possibly because the individuals seldom have the opportunity to see themselves in relation to the whole situation of which they are part. Nonetheless, the majority of picture development and model construction is usually undertaken without direct involvement of the client, who may be a busy director or department head. However, during the initial selection of relevant systems the client's opinion should be sought to avoid wasting time by pursuing inappropriate models.

The dangers of limited involvement are fairly obvious, ie that the analysts could devote a lot of effort to devising models that are not relevant, or become misled by issues or assumed facts about the organisation that are later proved to be incorrect. It is said that models are often developed intuitively, and this is frequently true; the generalised nature of high-level models makes this possible, but as the analysis proceeds to a lower and more detailed level, an input of specialised knowledge may become necessary. For example, it is easy, and valid, to assume that a notional system might have a component for *paying salaries*, but such intuitive assumptions are less credible when attempting to identify the associated activities, such as calculating salary rates, checking staff attendance hours, keeping accounts, etc.

It is, of course, a matter of balance. The majority of systems work may of necessity be carried out in the background, but at certain stages the analyst must check back with the client. Although a variety of different and seemingly relevant ideas about systems could be pursued in isolation, at the stage where one or more are considered suitable for a detailed modelling exercise, agreement needs to be obtained before proceeding. Experience has shown that even infrequent discussions, if held when the analysts have clarified their own thoughts about the situation, (ie not prematurely), can keep a project on the right lines, and prevent too much nugatory effort. Such discussions should be preceded by a presentation on systems concepts, tailored to give the minimum information needed for the client to understand the language of the analyst, and the approach being taken.

If a member of the client organisation is attached to a project team, there is a risk that the project development will be unduly influenced by an individual who could have entrenched views about the organisation. It is easy to voice the usual platitudes about the integrity and objectivity of analysts, but it is unrealistic to expect this to be maintained in all cases. There can be significant advantages in having an expert involved in a systems review, provided a balance can be achieved on the subjective/objective dilemma, which once again can be aided by the involvement of the main client at various stages.

Account must also be taken of the role of the full-time internal consultant, who may have worked for many years for the same organisation, and who is an integral part of the situation under examination. The knowledge of such consultants shouldn't be undervalued, and can reduce the number of discussions needed with the client. This is also relevant at the stage where feasible and desirable changes are being considered. The internal consultant should be aware of the prevailing climate, eg political, financial, sociological, and be in a position to discard certain potential courses of action that are, as a consequence, not worth pursuing.

Overall, the level of client involvement will depend on individual circumstances. The client may be willing to have a day-to-day exposure to the project, and have some working knowledge of the systems approach or be willing to learn about it. At the other extreme, the analysts may be left to their own devices, and only able to provide some education of their ideas, and confirmation of findings, when interviewing staff for fact-finding purposes. In any event, it is necessary to tailor the approach to suit the circumstances and progress the study accordingly. In the majority of cases, acceptable results are obtained even with limited client/analyst dialogue, provided the analysts are able to judge when such dialogue is essential (eg when selecting relevant systems, or focussing on particular problem areas), and make the most of the opportunities that arise.

8.5 Conclusion

In the preceding chapters, each stage of the SSM has been explained in detail, with some elementary examples given of how it might be applied in practice. A number of points have been deliberately emphasised throughout, such as the participative nature of the methodology, the lack of a precise sequence of actions in practice, and the need to be flexible when using the approach. This latter point is probably the most important, as, although the suggestion might be considered heretical, in the real world of the full time analyst or consultant, the end product is more important than the precise compliance with procedures or dogma. It has has already stated that the SSM provides *guidelines* rather than a *prescription;* with this in mind, the following chapters endeavour to illustrate ways and means of taking the essential principles of the soft systems approach and applying them to good effect in practice.

9 Practical Applications

9.1 Introduction

At this point in the book, the reader who has covered all the preceding chapters might be a bit overwhelmed by the constant references to systems theory and the SSM, and have some difficulty in making the connection with day-to-day analysis work. At first glance, the approach can appear extremely tortuous and consequently only suited to major investigations that aim to reveal fundamental problems within organisations, and where it is considered worthwhile to make an extra intellectual effort. I have often had this reaction from would-be users, which is one of the reasons why the book is entitled *Practical Soft Systems Analysis;* first, to convey the impression that it deals with making effective *use* of soft systems ideas, and secondly to avoid the suggestion that it is only concerned with the SSM. Whilst acknowledging that the ideas expressed in the book are based on those developed and fine-tuned by exponents of this methodology, these ideas can be of practical value when used on their own for general analysis work.

This chapter, therefore, is primarily concerned with putting soft systems analysis into the context of everyday practice, covering first some applications of general interest, then moving on to consider their use for management services work, and finally exploring how they relate to computer systems analysis. Whereas this distinction between the management services and systems analysis roles is somewhat artificial nowadays, it still provides a convenient baseline for readers who wish to consider the approach in terms of their own specialisation. The chapter also serves to introduce the final sections of the book, which cover the employment of soft systems ideas in greater detail.

9.2 General Applications

Taking the view that the SSM provides guidelines for *identifying* problems rather than *solving* them leads to the conclusion that it would not be applicable if the problem is clear and well-defined, and more scientific techniques would be used instead. This is true in relation to the methodology as a whole, which is essentially about exploring a situation to enable a dialogue to be established with the client about ways of making improvements. Once this dialogue is effective, and more discrete problems identified, then other techniques are required to develop solutions. At times, of course, what appears to be a clear unequivocal study of a specific problem turns out to be one that requires better definition, in which case the SSM could once again prove useful (Fig 9.1).

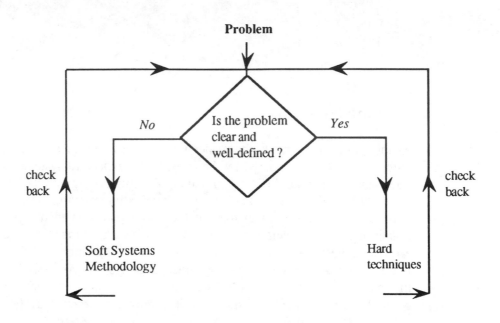

Fig 9.1 - Use of the Soft or Hard Approaches

However, this simplistic view can be misleading, as systems ideas generally can be of value in a variety of situations, even where the brief is explicit, and the problem well defined. To explain this, it is first necessary to consider how the SSM differs from other more conventional approaches, and whether these differences are significant in practical terms. Stages one and two encompass activities that are very similar to those carried out at the start of any study, ie using appropriate techniques to uncover the relevant facts about the situation. A rich picture is a novel way of illustrating these factors and, whether used as a basis for selecting relevant system viewpoints or simply as an aide memoire, is a device that many analysts will find helpful. However, the output from the early stages of any study will be an *expression* of the pertinent details, conventionally as notes, tables, reports etc, but in essence achieving the same ends as the summary given in a picture. Bearing in mind that graphic illustrations are frequently used by most analysts (eg for flowcharting etc), the concept is not entirely new, although a rich picture has an unusual style, and is less structured than other diagrammatic techniques.

Leaving aside the systems thinking exercises for the moment, the comparison stage involves activities in the real world that are not unlike those of conventional approaches, that is, posing pertinent questions about such matters as organisation structures, efficiency, effectiveness, procedures and so on. Likewise, stages six and seven (ie selecting feasible and desirable changes/taking action to improve) are invariably undertaken in some form as a result of a study or review. The SSM is applied at a higher level with a fundamentally different intent, ie to help comprehend a complex situation and identify potential improvements, rather than addressing and solving specific problems. Nonetheless, the actions that would be taken and the points to be considered, such as changes in structure, procedures, and attitudes, feature in most analytical work, and are generally covered in the training that all analysts receive, whatever their specialisation.

It can be argued, therefore, that many aspects of the SSM involve activities that take place in many studies, and it is only the *intention* and the *sequence* of these activities that are essentially different. However, the practice of constructing root definitions and conceptual models represents a significant change in approach, one that positively encourages lateral thinking and the development of new ideas. It is generally accepted that this part of the SSM is most closely aligned to a technique, and this technique can be used on its own across a broad range of analytical work. Taking a systems view will not only facilitate learning as ideas are generated, but also encourage a more structured approach to investigative work. The link between the two adjectives **systemic** and **systematic** can be recognised here; the view taken of human activity is a **systemic** one (ie regarding it as a system), which can result in the development of a **systematic** (ie methodical) approach to problem solving.

These are fairly broad statements, but they are valid because all human activity can be viewed in systems terms, eg the activities of an organisation, planning and organising a project, putting a man on Mars, taking the dog for a walk, and so on. It is interesting and entertaining to consider familiar situations in this way, for instance, what sort of system does a marriage represent? Is it one for the procreation of children, for establishing some order in a society, for satisfying baser human instincts, and so on. Modelling these ideas to determine the sub-systems, activities and measures of performance can be quite revealing, and might lead to a better understanding of why some marriages appear to work better than others, and why so many divorces occur. Similar exploration into the worlds of politics, economics, entertainment, sport, etc can be equally fruitful, and exponents can become so obsessed by this notion that they expound it with a religious fervour (which could be one of the causes of divorce amongst the soft systems fraternity !).

Without going this far, most analysts will find it useful to think in system terms about the work that they do. This doesn't mean that formal exercises in developing root definitions and conceptual models are always necessary. Considering human activity in terms of inputs, transformations and outputs, and keeping in mind the basic systems template (Fig 9.2) can help in the development of ideas about situations, and methodical ways of approaching them. This is demonstrated by the Brian Wilson *model for implementation* shown in Chapter 8, which helps to structure the approach to the execution of changes in an organisation.

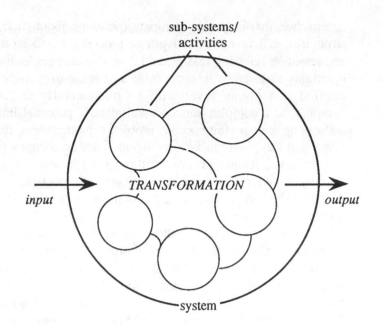

Fig 9.2 - Basic System Template

To illustrate the value of thinking in systems terms, a variety of examples are given in this chapter which reflect the generic nature of the phrase **soft systems analysis**, ie referring to all analytical activities that employ systems ideas to a lesser or greater extent. At the same time, it is helpful to differentiate between the type of studies where the SSM as a whole could be applied beneficially, and those where even a limited amount of systems thinking will be of value.

9.2.1 Loose or Tight Briefs

Investigations can be broadly divided into those where the brief is a loose one, (such as *to review a department, to improve value for money, to advise on an IT strategy,* etc), and those where the problem has been specified by the client, and the brief is fairly tight (eg *to advise on equipment purchases, reduce staffing levels, design an office layout,* and so on). Loose briefs generally apply when the client has a vague feeling of uneasiness that things are not as they could or should be, and the analyst has to advise on **what** needs to be done rather than just **how** to do it. In this case, there is also a need for a high degree of client participation, particularly when developing ideas about problems and possible solutions. These type of studies are obvious candidates for the SSM.

Tight briefs, however, are those given where the investigation is addressing the problem of **how** to achieve a desired solution, and the result is predictable within certain parameters. Client involvement will be limited to the initial discussions when establishing terms of reference, and during the final stages when recommendations have been formulated by the analyst. In this instance, the analysis can often be enhanced by a limited amount of systems thinking, using the basic template as a guideline.

Examples of typical applications within these **loose** or **tight** categories are shown in Fig 9.3.

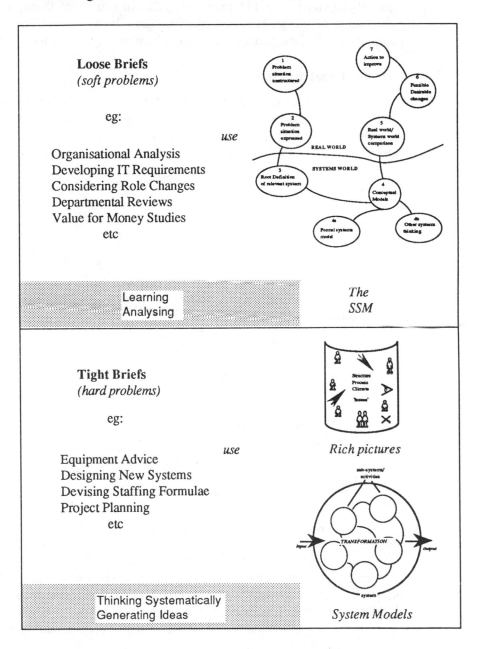

Loose Briefs
(soft problems)

eg:

use

Organisational Analysis
Developing IT Requirements
Considering Role Changes
Departmental Reviews
Value for Money Studies
etc

Learning
Analysing

The SSM

Tight Briefs
(hard problems)

eg:

use

Equipment Advice
Designing New Systems
Devising Staffing Formulae
Project Planning
etc

Rich pictures

Thinking Systematically
Generating Ideas

System Models

Fig 9.3 - Summary of Possible Application Areas

9.2.2 Loose Briefs - Use of the SSM

The loose category has been considered at length during the chapters that explain the various stages of the SSM, and in practice it has proved invaluable for revealing structural problems, parts of the organisation that have simply absorbed new requirements without explicitly adapting to meet them, and where the expectations of clients have not been recognised and allowed for. The following paragraphs give some further examples, which are expanded in later chapters.

Organisational Reviews

Fig 9.4 illustrates a typical SSM application, eg for an organisational/departmental review. This is carried out by designing a notional system model that reflects the primary task of the organisation in question, then comparing the model with what is happening in practice, indicating where improvements are possible.

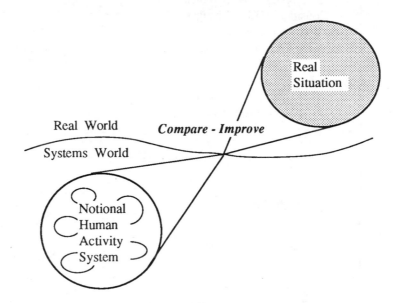

Fig 9.4 - SSM for Organisational Analysis

As well as looking at the formal structures and functions that exist, during studies such as these it is worthwhile to consider what systems they represent, and for what purpose they are established. Frequently this underlying purpose isn't apparent, leading to inconsistency of service levels, inefficiencies, duplication of effort etc; more importantly perhaps, causing dissatisfaction when the role of an individual in relation to the work of the organisation as a whole is not clear. Systems thinking can draw out the commonality of purpose and identify values that could be shared at all levels.

Determining Requirements for New Technology

Essentially, organisational analysis is at the root of using SSM to advise on the requirements for new technology, as it encourages the analyst to consider the overall purpose of the organisation, develop root definitions and conceptual models to illustrate this purpose in systems terms; decompose and compare the models with the real situation; and, where there is a mismatch, consider the use of technology to achieve an improvement. In this way, the technology that is chosen supports the organisation's prime purpose, rather than simply mechanising particular functions. Using the SSM to supplement systems analysis techniques is a feature of a number of methodological packages, some of which are discussed later in this Chapter.

Considering Role Changes

In a similar manner, where a significant change of role is anticipated, a model can be developed that reflects the new requirement, which is then compared with a model that reflects the existing role (Fig 9.5). This approach was used to good effect when examining the future requirements of Colleges of Further Education in light of new legislation, which is covered in Chapter 12.

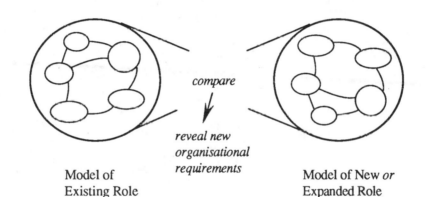

compare

reveal new organisational requirements

Model of
Existing Role

Model of New *or*
Expanded Role

Fig 9.5 - Considering Role Changes

9.2.3 Tight Briefs - Thinking in Systems Terms

Tight briefs probably arise more frequently than others, and it is in these areas of work that selected elements of the SSM can help on a day-to-day basis. Whilst sitting at a desk wading through a stack of background papers, it is worth noting down in picture form all the significant points that arise, then thinking about the inputs, transformations, and outputs that are needed for a particular operation to function effectively. Normally this doesn't take very long, but it is surprising how it can help with collecting thoughts and deciding the approach to be taken. Some examples of using systems ideas in this manner are given below.

Developing a Staffing Formula

If there is a requirement to set up a staffing formula (ie a means of reviewing staffing levels on the basis of known work generators), systems thinking can help to clarify and separate the activities of an operator into discrete elements of work, which can then be measured to derive the factors for the formula.

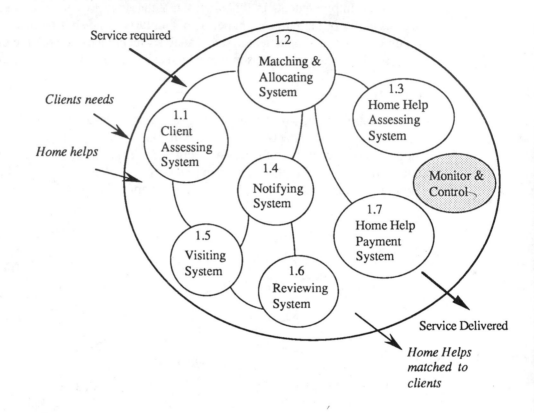

Fig 9.6 - Derivation of a Staffing Formula

Fig 9.6 illustrates how this was achieved when examining the activities of Home Help Organisers, whose role is to assess the needs of individual clients, consider the skills and availability of Home Helps in the same geographical area, and ensure that a good match is made. Each of the sub-systems of the model was regarded as a workload generator, and the formula was derived by measuring the time for each element, then multiplying by the number of occurrences each year.

Developing a New System

If there is a totally new requirement, a model of the process needed to arrive at the desired state (ie with the new organisation installed and functioning) can be constructed, providing a checklist of the factors that need to be considered before the new operation can be established. For instance, setting up a records management function within an organisation will require such activities as defining the potential customers, quantifying the records to be held and accommodation needed, calculating staffing levels etc, as shown in Fig 9.7.

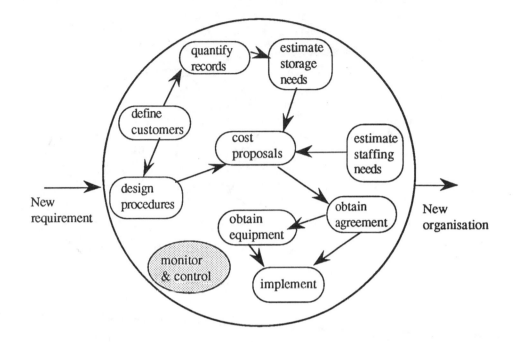

Fig 9.7 - Developing a Records Management System

Generating Ideas

Less specifically, a brief excursion into systems thinking can help with the generation of ideas about a wide range of subjects. When advising on equipment selection, it can be of value to view the items concerned in terms of the wider systems that are served; relationships and communications between individuals and groups can be explored *systemically* when developing office and building layouts; the wording of policy statements can be examined to determine if they are achievable, and if this achievement can be measured in some way, and so on. This doesn't mean that systems thinking provides a universal panacea to solve all manner of problems, but it can be used quickly and effectively to view situations in a detached and productive manner.

Project Planning

One final example is useful to demonstrate how it also helps to develop a methodical approach to such activities as project planning and control. The input in this case could be a request for the study, and the required output a study plan; leading to a systematic derivation of the activities needed before the transformation can be achieved. It is worth noting that this doesn't represent a description of the actual planning activities, although this could be derived by decomposing the model, but provides a high-level checklist of the main components of the project planning process (Fig 9.8).

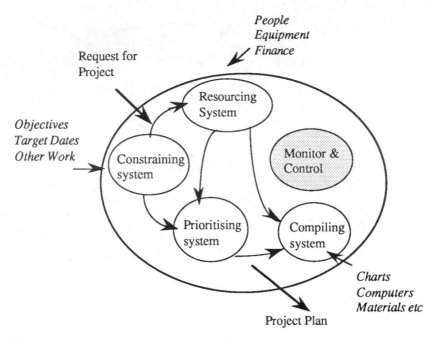

Fig 9. 8 - A Project Planning Process

9.2.4 Summary

It can be seen from these examples that the SSM and the application of systems ideas to typical situations can both be of benefit in a variety of ways. Certain general conclusions can also be drawn about its use by the problem-solving professions, in the two areas of management services and systems analysis, and these are discussed in the remainder of this chapter. Additionally, systems thinking should not be regarded as exclusive to full-time analysts; the term is used throughout the book to mean *those who wish to analyse*, and applies to any individuals who would like to learn more about their own situations. This could include managers seeking some improvement to their operations, salesmen reviewing their sales techniques, administrators aiming to improve clerical procedures, and so on.

9.3 Management Services Applications

It is difficult, and unwise, to be prescriptive about the use of any techniques or methodologies for investigative work; the decision about which tools to use will always depend on the circumstances that prevail. Likewise, it is not possible to specify whether systems thinking is more suitable for O&M, Work Study, Systems Analysis etc; in other words, to *slot* the approach into a particular compartment of the management services toolbox. However, bearing in mind the examples given where the full SSM has been used successfully, and those where systems ideas alone have been of value, it can be concluded that the approach is more suitable for work carried out under the broad heading of *organisation and methods*. It has some application in the area of Work Study, but is limited to clarifying the context of the studies and encouraging a systematic style.

O&M, however, has developed over the years from parochial studies concerned with improving clerical and administrative procedures to a general problem-solving service for management, implying that, as the problems facing management change, then the skills of the practitioner need to evolve to reflect that change. This has been the case in recent years with the involvement of O&M officers in the introduction of computer systems, from the definition of user requirements to the implementation and support of microcomputers and wordprocessors. Apart from those practitioners that have chosen to specialise in one of the nested disciplines, the work is now fairly broad-based, covering the whole spectrum from structural reviews to business systems analysis. It would seem, therefore, that if soft systems analysis has to be *slotted in* to the management services toolbox, then it fits most conveniently into the O&M compartment.

Questions are often asked about the comparison between this form of analysis and the traditional *techniques* used in the profession. This comparison is not really valid as the level of analysis is different, and the soft systems approach provides guidelines rather than the strict procedures that the term technique implies. However, there are certain parallels that can be drawn between these guidelines and some that are used for O&M work. The **SREDIM** mnemonic, for example, provides a simple guide to the typical stages of a project, ie:

- *S*elect
- *R*ecord
- *E*xamine
- *D*evelop
- *I*mplement
- *M*aintain

Whilst it is possible to undertake some or all of these activities in sequence, in reality most of them are happening simultaneously, and there is a large amount of iteration when moving through the *examination* and *development* stages. The point was made earlier that many of the activities undertaken during a systems review are comparable with those of other approaches, and these similarities can be recognised by viewing the stages of SREDIM alongside those of the SSM (Fig 9.9). This isn't meant to suggest that the actions taken in all cases are the same, but that the type of thought processes involved are similar, likewise the required outputs from the various stages.

In particular, the diagram shows the *selection* stage being common to both, but this may well take place at different levels, ie for an O&M study the problem could be well-defined, whereas with the SSM, only the *problem situation* will have been chosen.

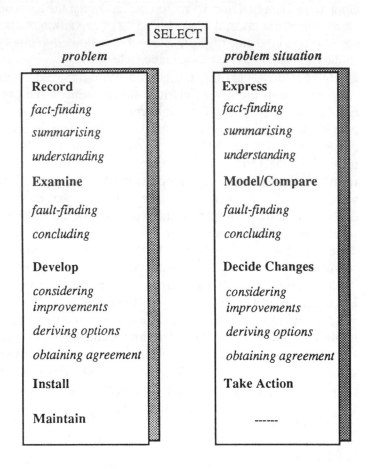

SELECT	
problem	*problem situation*
Record	**Express**
fact-finding	*fact-finding*
summarising	*summarising*
understanding	*understanding*
Examine	**Model/Compare**
fault-finding	*fault-finding*
concluding	*concluding*
Develop	**Decide Changes**
considering improvements	*considering improvements*
deriving options	*deriving options*
obtaining agreement	*obtaining agreement*
Install	**Take Action**
Maintain	------

Fig 9.9 - Comparison of SREDIM with the SSM Stages

9.3.1 Other Application Areas

Detailed examination of the use of soft systems analysis for other types of management services work is possible, but would probably be misleading as each study has different characteristics. Returning to the toolbox analogy, systems thinking is an *adjustable spanner* which needs to be altered to suit the variety of *nuts* that can be found; the fundamental principles remain the same, but modifications are required to suit each particular situation. Nonetheless, some general guidance can be found from the various points raised in this chapter and those considered elsewhere in the book, as summarised in tabular form in Fig 9.10.

MS WORK AREA	SOFT SYSTEMS ANALYSIS INPUT	REFERENCE
Method Study	*Procedure audit*	Chapter 13
Value for Money Reviews	*Organisational analysis*	Chapter 10
Productivity Schemes	*Planning, organisational analysis*	Chapters 9, 10
Work Study	*Planning, generating ideas*	Chapter 9
Procedural Reviews	*Procedure audit*	Chapter 13
Department Reviews	*Organisational analysis*	Chapters 10,12
Staffing Issues	*Derivation of activities*	Chapter 9
Equipment Selection	*Putting into context, generating ideas*	Chapter 9
User Specifications	*Organisational analysis, information enquiry*	Chapters 10, 11
Business Analysis	*Organisational analysis*	Chapter 10
Layout Design	*Logical relationships*	Chapter 9

Fig 9.10 - Examples of MS Applications

9.4 Computer Investigations

Soft systems analysis does not specifically address the development of computer system specifications, and consequently doesn't replace or preclude any other systems analysis and design methods. It is often used as a prelude or supplement to these methods, mainly to give a well rounded view of situations as the technical analysis proceeds. Some investigative packages, notably the FAOR approach, also adopt some of the notions used with the SSM to consider computer developments in a variety of ways, eg as an information system, a communication system, or a functional support system, etc. To illustrate these types of applications, a brief description follows of three such packages, ie, **FAOR, MULTIVIEW** and **COMPACT**, with more detailed information available from the sources listed in the bibliography.

9.4.1 The FAOR Methodology

The FAOR (Functional Analysis of Office Requirements) methodology is a comprehensive approach for determining the requirements for office support systems. In brief, several components are drawn together under a loosely procedural framework of analytical activities aimed at identifying the requirements for an office support system, as part of the overall development life-cycle. The package provides the analyst with an effective means for exploring a given situation, and for focussing on those areas where technology could significantly improve the achievement of selected organisational objectives. (*Functional Analysis of Office Requirements - A Multiperspective Approach*, by G Schafer, published by John Wiley and Sons, 1988)

It views an office as a system where human activity is taking place, and where the activity is largely unstructured and ill-defined. The analysis is carried out as a series of activities (A1 to A4 in Fig 9.11), moving from the preparation of a study brief in consultation with the client to the selection of suitable *instruments* for a detailed examination of selected areas. A detailed statement of system requirements is then produced and the likely benefits and disadvantages of alternative computer solutions evaluated.

The SSM is used in three contexts. During the Office Exploration (A1) it assists the analyst in understanding the client organisation and focussing on problem areas. During the Method Tailoring exercise (A2) it assists in constructing a method from the 'instruments' according to the defined study objectives. In the Requirements Analysis (A3) it provides a basis for coordinating the results of the previous study work, and controlling the progress of the study towards a fulfilment of the study objectives. A detailed discussion of the FAOR methodology is contained in Chapter 14.

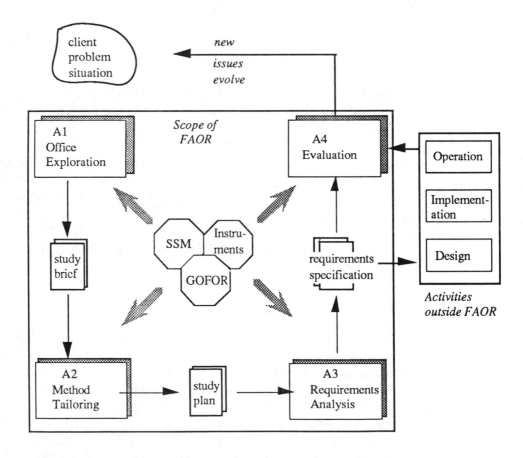

Fig 9.11 - The FAOR Methodology

9.4.2 The Multiview Approach

The Multiview package, as described in the publication *Information Systems Definition: The Multiview Approach* (Wood-Harper, Antill and Avison, Blackwell Scientific Publications Ltd, 1985), recognises the need to consider the functional, technical and sociological aspects of an organisation when carrying out a computer systems investigation.

The approach has five main stages as shown in Fig 9.12 (based on the diagram given in the Multiview publication), which can be summarised as:

1. The analysis of *human activity* systems.

2. The analysis of entities, functions and events, referred to as *formation modelling*.

3. The analysis and design of a *socio-technical* system.

4. The design of the *human-computer interface*.

5. The design of the *technical subsystems*.

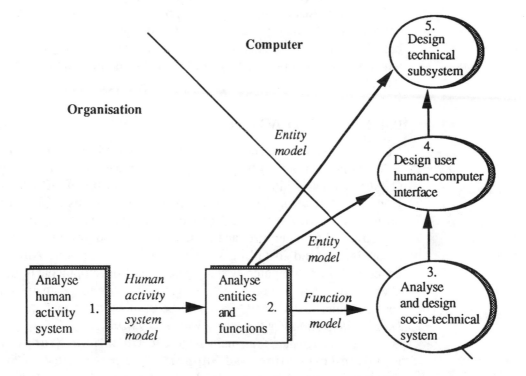

Fig 9.12 -The Multiview Approach

In this methodology, the soft systems approach is combined with more conventional techniques (eg data analysis and structured analysis) to create a framework for the computer system design which takes account of the viewpoints of all the people who will be involved. For the technical analysis and design aspects the techniques of Structured Systems Analysis and Design are drawn on, together with the standards developed by the National Computer Centre.

Socio-technical issues are addressed using participative design methods such as those developed by E Mumford and R Hirscheim, and human-computer interface applications. Of particular interest is the use in the early stages of the soft systems ideas to examine the organisation and draw out, in consultation with the client, statements about purpose of the organisation and a clear idea of what the information system will be in functional terms, and what it will do.

The Multiview approach quite deliberately moves from generalisations about the organisation to specific details, the outputs from each stage becoming inputs to the following stages, or major outputs of the methodology. Soft systems ideas are used to structure a debate and subsequently identify problem themes, eg conflict between departments, poor communications, lack of co-ordination, and so on. These problem themes are then examined to determine if and where technology can be used to improve the situation.

The progressive movement from an examination of the organisational objectives and related problems to technical (and other) solutions ensures that the technology is used to improve the performance of the organisation in relation to its defined role. At the same time, the detailed, non-technical, investigation covering both functional and human aspects gives the analyst far greater insight into the whole situation than would be gained by using computer systems analysis methods only.

9.4.3 The COMPACT Approach

COMPACT is a comprehensive package developed by the Central Computer and Telecommunications Agency of the Civil Service, covered by Crown Copyright (1986), with the aim of improving the operation of offices within central government. It recognises the value of the soft systems approach for use where problems are ill-defined and office activities may not support the business objectives of an enterprise, and these aspects are addressed as part of the process of designing and installing systems for office support. Although primarily concerned with computer systems design and development, it also accepts that alternative (ie non-computer) solutions may result from the analysis, and makes the specific point that COMPACT could be used by managers and staff *"to gain a better understanding of the area in which they work"*.

The package has four main stages, ie **Initiation, Study, Development and Implementation,** and **Support.** The principal use of the SSM occurs during the Study Stage (Fig 9.13) in a modified form developed by Systems Concepts Ltd. This is referred to as a *"Business Analysis Methodology"*, which aims to understand and solve business problems, or exploit business opportunities. During the business analysis phase, a formal definition is prepared that reflects a particular perspective of the situation, stating what the business aims to achieve, what its components are, and how they interact. From this definition a model is prepared, which is then expanded until the information requirements can be identified. This *information analysis* exercise (similar to that described in Chapter 11) is then used as the basis for a new office system, or for comparison with existing practices during the *organisational analysis* phase. As a result of these examinations a preferred business solution is prepared, ie one that defines the desired activities and standards to satisfy the objectives of the business, the needs of staff, and any technical constraints.

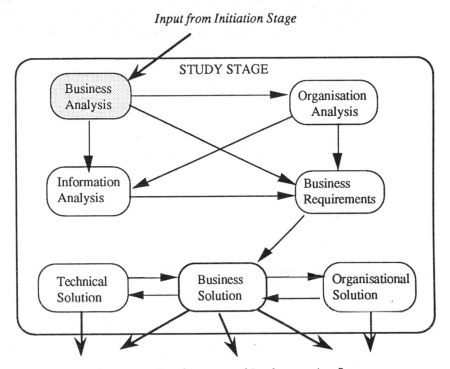

Input from Initiation Stage

STUDY STAGE

Outputs to Development and Implementation Stage

Fig 9.14 - Study Stage of the COMPACT Approach

Based on this statement of requirements, the technical implications are considered, together with the impact on the way in which work is organised. The outputs from the Study stage are then used as the basis for technical design, equipment procurement, and installation in the next stage, with support services and follow-up evaluation action being considered during the final phase.

SSM or an associated derivative is essentially used in all three of these packages in the same way, ie to put computer developments into the context of the whole situation, not only taking account of technical requirements, but also those of the people involved, and the total business requirements of the enterprise being examined.

9.5 Conclusion

In the course of this chapter, a variety of uses of soft systems ideas have been explored, some contained within packaged methodologies, and others based on informal application for general analysis work. It must be clear by now that, although the approach seldom solves problems, it helps to understand them and address their solution in an ordered and logical manner. It must also be apparent that the only real limitation on the use of systems ideas is the imagination and initiative of the analyst, a point which is made abundantly clear in the final chapters of the book that deal with typical problem-solving investigations.

10 SSM for Organisational Analysis

10.1 Introduction

The next two chapters contain details of research projects that were carried out to test the use of soft systems ideas in typical problem situations. The project described in this chapter, which was concerned at the outset with identifying where new technology could be used beneficially, also provides an excellent example of the value of the SSM for learning about an organisation and its weaknesses. It demonstrates that the main output from the exercise is improved knowledge, knowledge which can then be instrumental in ensuring that any changes proposed by the analyst, including new technology developments, take account of the whole situation and support the client's primary objectives. The project addresses the Social Services department of a County Council in the United Kingdom; however, the conclusions in terms of the value of the approach are equally valid in any other organisational setting.

10.1.1 Background

By way of background information, local government in the United Kingdom is exercised at three complementary levels, ie by Parish Councils concerned with the local interests of small areas or *parishes*; by District Councils responsible amongst other things for the maintenance of local roads and the collection of refuse within wider *district* boundaries; and by County Councils covering the largest administrative and judicial areas into which the United Kingdom is divided.

A County Council provides a wide variety of services, ranging from the maintenance of the major highways and motorways that pass through the area, to the provision of education and social services. The work of the Council is democratically controlled by elected council members who sit on a number of specialist committees, eg Education, Highways, Planning, Social Services etc. These committees are supported by specialist departments, each with its own Chief Officer who has delegated functional responsibilities for implementing the decisions of the committees, and of the Council as a whole.

10.1.2 Context of the Study - The FAOR Project

The study formed part of a larger research project carried out within the framework of the European Programme for Research into Information Technology (ESPRIT). This Programme, sponsored and part-funded by the European Economic Community, aims to bring together in constructive partnerships academic, research and commercial organisations throughout Europe, with common interests in certain aspects of information technology development. By pooling resources at the pre-competitive stage of development, all participants can benefit by having access to a wide range of facilities, experience, and intellectual ability. The main partners responsible for a nominated ESPRIT project have the authority to sub-contract elements of the project as required, thus allowing an even wider range of organisations to be involved in the research.

The ESPRIT Project described here and elsewhere in the book was directed at the development of a methodology for the Functional Analysis of Office Requirements (FAOR), and the full results of the study are contained in the FAOR publication referred to in Chapters 8 and 9. In brief, the approach identifies and evaluates the requirements which determine the design of office systems, ensuring that they support the achievement of selected organisational objectives. The SSM is used within this framework to analyse and describe office activities, and to structure the analysis activity itself.

The FAOR project commenced in 1984, and was completed in June 1987. In the early stages of the project, although it was accepted that the SSM seemed appropriate for inclusion in the final package, the project team had little experience of its application in practice. Accordingly, agreement was obtained to carry out a trial of the methodology at a County Council in the United Kingdom, with the prime aim of assessing the suitability of using the SSM as part of an investigation into information technology needs. This *Field Trial* was the basis for the case study described in the following pages.

In the latter stages of the FAOR Project, when the full methodology was suitable for testing, the SSM was again used beneficially to develop ideas about an improved 'information provision system' to support the objectives of the organisation under review, as covered in Chapter 14.

10.1.3 Approach to the Study

With the benefit of hindsight, it would be easy to imply that the Field Trial progressed in a fairly logical manner, and that each stage of the process was well thought through, and achieved benefits for the project in an efficient manner. The reality was far different, with the main value to all participants being a thorough grounding in the application of soft systems ideas to a practical situation.

From the beginning there were the inevitable logistical problems that arise when project team members are located on different sites, in this case spread across the face of Europe, in Denmark and Germany as well as the United Kingdom. In the early stages, a great deal of time and energy was directed at establishing working relationships, and laying down the principles and practices for communication and liaison. This obviously entailed a number of meetings and workshops at sites in Europe, and, although the visits were enjoyable, struggling through baggage checks at international airports just to clarify some minor administrative or technical point can be frustrating, and have a distracting effect on the serious and concentrated thought processes that are necessary to progress a research project.

At the outset of the study there were few guidelines available on applying the SSM, and, although the project team had read the main publications describing the approach (ie Checkland and Wilson), none of them had used it in practice. As a result, the initial work was carried out in a laborious, time-consuming manner, with a great deal of soul searching and intense discussions about what the project was aiming to achieve and the manner of achieving it. Even the unfamiliar terminology caused problems, and agreeing a consistent definition of such phrases as 'problem situations, primary tasks' and so on led to further time-consuming deliberations and subsequent lack of progress.

In the circumstances, it was agreed that the approach would be a heuristic one, testing out systems ideas as the study progressed, then refining the approach as the team's knowledge and confidence increased. In parallel with the SSM activities, it was therefore necessary to utilise conventional methods for summarising and reviewing the study findings, such as textual descriptions based on interview notes, and organisation charts, tables and so on.

In the event this hybrid approach proved more difficult than anticipated, and no real progress was made until the team attended a training session on the SSM given by a consultant from the South Bank Polytechnic of London. The training session provided a forum for discussion of soft systems concepts, and the opportunity to stand back from the complex real-world situation under examination and apply some *systems thinking* to it. As a result, the team were able to build on the knowledge they had already gained about the client organisation, and start to view it in systems terms; from this point on ideas gradually emerged, not only about improvements that could be made, but also about the value of the SSM when used in this context.

The next few pages cover the progress of the ESPRIT Field Trial, and are based on a report rendered to the Project Leaders in June 1986, expanded where necessary for clarification. For ease of reading, the description is presented here in a fairly logical fashion, but should be viewed in light of the previous comments about the underlying difficulties in this particular project. Nothing was quite as simple and straightforward as it is made to appear !

10.2 The Client Organisation

The Field Trial took place in the Social Services Department of the County Council, in broad terms a function responsible for providing caring support for members of the community who require such assistance for all or part of their lives. The

Department was not chosen at random, but because the Director in charge had expressed concern that technology was not being used as it could or should be, without any clear idea of how it could be employed to better effect. Like all the Council departments at the time, there had been some investment in wordprocessors and microcomputers, and a service was provided by a mainframe computer for such things as time-sheet and salary processing, and other corporate 'number-crunching' operations.

The Director, however, was concerned that most of the existing applications were directed at administrative functions, and felt that technology could be put to better use in support of the front-line social services. Consequently, he agreed to participate in the research in the hope that it would help to clarify where future developments would be more advantageous in terms of the organisational objectives.

To avoid restricting the examination until it was possible to focus on specific problems, the draft Terms of Reference prepared for the study after the initial meetings were deliberately broad-based (ie a *loose brief* as discussed in Chapter 9). The study was directed at the branch of the Department that provided the front-line *operational* services (ie the Operations Branch), and the Terms of Reference were expressed as:

1. To assist in understanding the client's objectives, particularly those concerned with front-line operations.
2. To assist the client in understanding his tasks and related issues.
3. To determine where Information Technology can improve the client's performance in meeting his objectives.

10.2.1 The Study Phases

Before considering the phases of the study, it is worth restating a number of points made in earlier chapters. Firstly, the SSM isn't usually applied in a straight-forward, sequential fashion, and it is always necessary to return to each stage and amend or up-date the rich pictures and conceptual models as new facts come to light, or new viewpoints are considered. Secondly, in practice it is unrealistic to expect the client to be available at the precise moment that the analyst feels that a discussion is necessary; consequently the development of models etc may well have to be undertaken out of sequence to avoid wasting time until the further consultation is possible.

In the case of the Field Trial, however, two distinct phases can be identified, the first addressing the Operations Branch as a whole, and the second focussing on a particular functional group, agreed as a suitable area to take some *action to improve*. The description of the study activities that follows is divided accordingly to recognise these two main phases.

10.3 Application of the SSM - Phase I

During the first phase, the Team were unsure of the direction that the study would eventually take, and consequently endeavoured to explore all aspects of the Operations Branch to identify discrete problem areas. Although this gave rise to a number of problems due to the size of the organisation being addressed (as discussed below), it eventually produced benefits by ensuring that all the significant factors that affected the situation were taken into account as the study progressed.

10.3.1 Overview of the Client Organisation

It has to be said that the Team, as well as the client, had a *vague feeling of uneasiness* at this point, not only because of inexperience with the methodology, but also because of the size of the organisation being examined. More than 6500 people were employed at that time by the Department, covering a wide range of administrative and support functions as well as those delivering the front-line services to some 250,000 clients. These services included residential, domiciliary, day centre, and social work facilities for a number of client groups, eg the elderly, children, families and the mentally or physically handicapped (Fig 10.1).

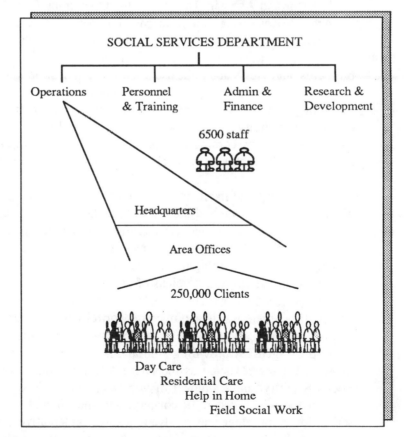

Fig 10.1 - Overview of the Client Organisation

The Operations Branch was divided into a Headquarters section, and a number of geographical divisions known as social work *Areas*, each managed by an Area Organiser with responsibilities for providing services to a specified region of the County. The Headquarters also included the supportive branches of Personnel and Training, Administration and Finance, and Research and Development.

10.3.2 The Problem Situation

The complexity of this extremely rich problem situation led to the Team's first dilemma, ie who, from a cast of more than three thousand actors, should be selected for interview so that relevant issues about the organisation could be identified and pursued ? There was little guidance on this from the existing textbooks, and the opinions of the team ranged from two or three key individuals to someone from each functional group, which could have amounted to more than a hundred persons. Eventually it was agreed that, as the Operations Branch was broadly sub-divided into Headquarters and Area functions, a reasonable cross-section of views would be obtained if representatives from each of these sub-divisions were to be interviewed. A selection was made of 27 interviewees representing the administrative and control functions at the Headquarters, including the Director and Assistant Director of the Operations Branch, and those persons at the Area Offices responsible for each client group. The interviews were carried out in a semi-structured manner, with each analyst working to a pre-prepared list of questions, primarily addressed at clarifying the role of each group, associated problems, and views about the use of computers in support of that role.

Even from this limited range of interviewees, over four hundred seemingly relevant points were raised; some that were minor grumbles about working conditions, scarcity of resources, and so on, and others indicating fundamental problems, such as communications within the Branch, co-ordination between groups, and the lack of any precise mechanism for assessing the effect of policy changes.

Bearing in mind the team's lack of experience of the SSM, it was at that time difficult to decide which of these should be treated as 'issues' in the SSM context, and which could be disregarded. Attempts to include every key point in the early rich pictures resulted in the pictures becoming too muddled to make sense of, although they served to highlight the complexity of the situation. Needs being the mother of invention, the problem of analysing the interview findings eventually resulted in the development of the *interview analysis* technique (as described in Chapter 5), which has since proved its worth in many other studies.

In these early stages, the team first encountered the dilemma caused by the overlap of agencies providing similar services to that of the organisation being examined, agencies such as the District Health Authorities, the Department of Social Security, and a number of voluntary bodies. These were considered valid parts of the wider system of concern, and the dilemma of how to recognise the separation at this stage was resolved by showing tentative system boundaries in the rich picture. Mixing systems ideas with facts about the real world may appear to

se as handout - for example??

'break the rules', but in this particular exercise it helped to clarify what was happening in practice. Eventually it was possible to compose a picture that captured some of the richness of the situation, shown in abbreviated form in Fig 10.2.

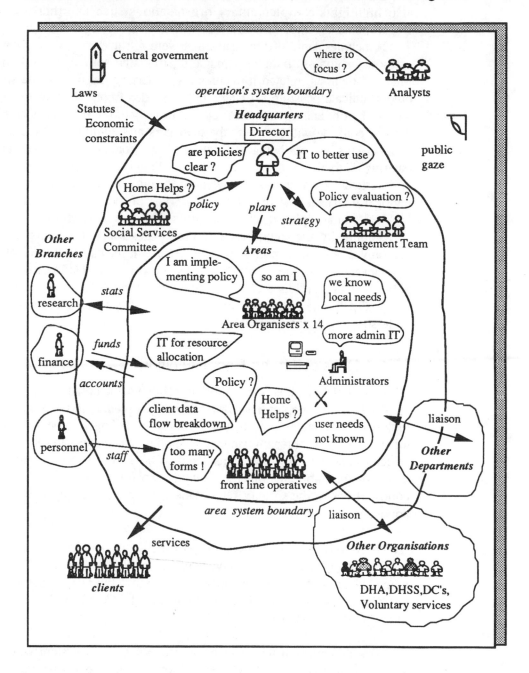

Fig 10.2 - Rich Picture of the Operations Situation

The picture reflects the structure of the Branch, showing the split to Headquarters and Area functions, and the other Social Services branches (Research and Development, Administration and Finance, and Personnel and Training). It also highlights complementary organisations such as other County Council Departments, together with government and voluntary agencies.

It includes some of the issues that were felt to be relevant, shown in typical cartoon fashion as 'bubbles' emanating from the source that raised the matter. Many of them are related to technology, reflecting the bias of the study, whereas others indicate possible problems on a broader front. There were signs that the views of front-line operators in relation to technology had not been clarified, and did not coincide with those of the managers and administrators; that information about clients was not being adequately shared; some concern about the amount of paperwork involved when processing a client; and a commonly held view that policy statements were not always clear, leading to different interpretations by the regional managers and front-line staff. Some of the areas of potential conflict are highlighted in the picture using the 'crossed swords' convention.

Admittedly, the significance of these issues was not obvious at this stage, nor was any one issue consciously selected as the basis for the development of a relevant system model. With hindsight, however, it was obvious from the root definition and models that were eventually constructed that the team had been influenced by the concern expressed at all levels about the communication and understanding of policy matters.

10.3.3 Developing the Root Definition

After much debate about the CATWOE elements, the broader implications of which are discussed in Chapter 6, two root definitions seemed to be appropriate. The first took a rather clinical view of the situation, regarding it generally as a system to manage given resources to implement policy statements, and to do so in a cost-effective manner. This was considered to be a W that represented the views of many of the public within the County, expecting value for money from their contributions to the County rates. This definition, with the CATWOE elements annotated, was expressed as:

'A system owned by the Director (O) of the Social Services Department, operated by professional and administrative support staff of the Department (A), to manage given resources effectively and efficiently (T) to deliver caring services to clients of the Department (C), within the constraints of the County Policy for the provision of care in the County (E)'

This definition of the notional *service management* system was considered to reflect the primary task of the organisation, with the level and standard of service clearly prescribed and constrained by County Policy statements. There is no mention of any requirement to clarify the needs of the clients, or to identify the clients themselves. The definition was not considered perfect by the team; in particular the assumed environment would impose constraints on the system rather than just influence it, but it was considered adequate to allow the study to progress.

The second root definition, although retaining the same recognition of actors, clients, ownership, and environment, had a significantly different transformation:

'to identify and meet the needs of the less-able residents of the County'

This was considered to be a 'warmer' statement than the previous one, reflecting a more emotive view of the role of the operations system. It is essentially issue-based, ie there will always be some contention as to the type and quality of services needed by less-able members of the community, or indeed, whether or not these services should be controlled by a local authority. Whilst accepting the constraints of statutory regulations for the provision of caring services, that is the regulations governing what services must be provided by the Authority, the main purpose of this notional system was to identify those people in the County in need of care; decide what that care should consist of, then develop policies and activities to provide it. A research function would also be required, establishing in total a *go out and find and then heal* system, reflecting the assumed viewpoint of, for example, a social worker, relative, doctor, or other 'caring' persons.

10.3.4 Constructing the Conceptual Models

Although the two root definitions were superficially the same, the significance of the different transformations became apparent when the conceptual models were constructed. Fig 10.3 shows the model of the first, service management, root definition.

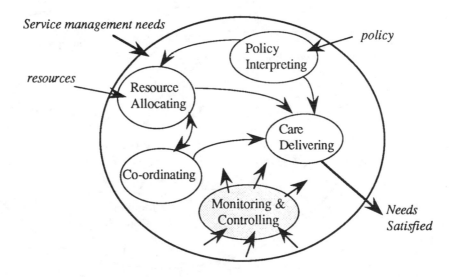

Fig 10.3 - Conceptual Model from First Root Definition

It is often said that initial attempts at conceptual modelling are undertaken intuitively, which implies that most persons with an analytical mind could reasonably work out what system components are necessary to achieve a stated transformation. Experience has shown that this is generally true, particularly when constructing high-level models, such as those considered relevant in the Field Trial. There is also an element of guesswork and a variety of models are constructed that appear to meet the requirements, which are then reviewed to determine whether or not they are relevant to the investigation. It seemed reasonable that a system to implement the requirements of policy should have components for *interpreting* this policy into action statements, *allocating* the given resources equitably, *delivering* the services and *coordinating* the associated activities, not forgetting the need to monitor and control the system as a whole. The influence of the earlier statements about policy can be identified by the inclusion of the *policy interpreting* sub-system.

The model constructed from the second definition contained the same basic components at its core, but was expanded to include activities needed to 'identify and meet the needs' of clients; for example *research* to clarify changes in demand, associated *policy-making* activities, and those required to *plan* and *allocate* resources as required. The result of this expansion produced a model that encompassed all social services functions in the County, obviously including many that would not come under the Operations Branch, and looking very similar to the Social Services Department as a whole ! Fig 10.4 shows the second conceptual model.

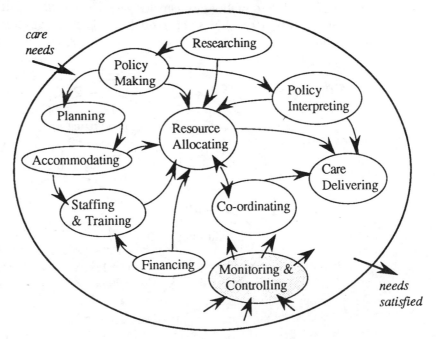

Fig 10.4 - Conceptual Model from Second Root Definition

This latter point also illustrates a further problem that can arise with soft system studies, that is the effect of 'real world' interference, when, despite the best attempts of the analyst to remain detached and objective, the thought processes are influenced by what is happening, or what exists, in practice. It is difficult to gauge how this affects the outcome of a study, as there is no yardstick to measure it against, but, provided ideas are still developed in an objective way, it needn't undermine the value of the systems thinking stages.

10.3.5 Making the Comparison

Even with hindsight, it is difficult to determine the exact point in the Field Trial when the modelling stage stopped, and the comparison with the real world started, emphasising that this is not a strictly sequential methodology. Ideas about problems were being formed and being discussed from the outset of the review, with the systems thinking exercise serving as a focussing mechanism to make these ideas more explicit. Neither was there any deliberate iteration; all the mental activities were going on concurrently, manifested by frequent changes to rich pictures and models as the study progressed.

A conscious attempt was made, however, to draw out from the models ideas about where problems might exist, and construct logical arguments as to why improvements appeared necessary, in other words to substantiate what the analyst and interviewees felt was wrong. For example, the model indicated the need for a 'policy interpreting' sub-system, prompted by the widespread concern over policy matters. To understand what effect the lack of related activities might have, and develop the rationale for making changes, the team decomposed this and the other elements of the model to the second level. First a root definition was drafted for each component, then a model drawn to show the activities necessary to achieve the transformation at that level and the associated outputs. A summary of the second-level models is shown in Fig 10.5.

This exercise indicated, amongst other things, that interpretation of high-level policy statements should produce clear directives about services for each client group, directives that would influence the way Area and establishment managers provided these services. They would also affect resource allocation, provide the basis for coordinating the activities of each regional division, and help ensure that a consistent standard of service was provided throughout the County. This seemed a fairly logical argument, implying that, if 'policy interpretation' was not taking place, then the remaining systemically desirable activities could not be effective. Furthermore, monitoring and control could not be achieved without firm guidance on the standards and practices expected from front-line establishments.

One of the dangers of thinking in systems terms is the likelihood of forming *ivory-tower* views about a situation, a point made in earlier chapters. Analysts can be so detached from reality that they become arrogant about the value of their ideas. In this instance, most of the views were based on remarks made by staff of the Operations Branch, and the modelling exercise put them into a systems context. There were, in fact, a limited amount of observable 'policy-interpreting' activities going on in relation to the services provided for children in the County. This highlighted by comparison those services that operated from somewhat vague policy statements, without clear directives as to what actions to take and the standards required.

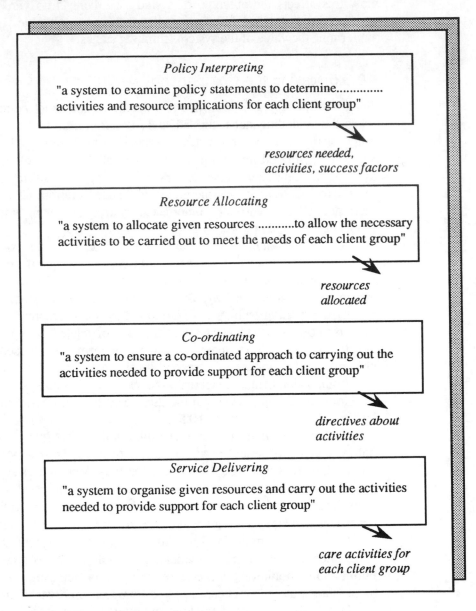

Fig 10.5 - Summary of Second- Level Modelling

The comparison stage, therefore, was not carried out by explicitly attempting to identify in the organisation those elements of the model that should theoretically exist, but by developing a logical argument about the cause of those problems bought to the attention of the team.

This argument led to the following conclusions:

1. There were only a limited number of activities concerned with interpreting policy statements into action statements, or with assessing the resource implications of policy changes. Given that policy was frequently stated in high-level terms, each region was free to interpret it differently, causing the standard of service to vary significantly from area to area.

2. Without some conversion into meaningful statements of what actions to take to implement policy, there could be no yardstick or critical success factors to measure and evaluate the effects. Therefore, fully efficient and effective service management could not be achieved.

3. Without specific guidelines about the actions to be taken for each client group, there could be no equitable basis for allocating resources to comply with policy. Furthermore, the branch had limited responsibility for deploying the total resources it was given - these were allocated to the front-line units by two other branches, ie the Personnel and Training branch, and the Administrative and Finance branch.

4. There was no formal system for coordinating the front-line activities. Any co-ordination that took place was through meetings of Area Organisers and other staff in an ad hoc manner. (Areas were in fact encouraged to develop their own style of operation as part of a decentralisation programme.)

5. The lack of any critical success factors made effective monitoring and control difficult. This was generally carried out in a semi-formal manner, for example through meetings of senior officers from the Headquarters and Area Offices, and by exception, when some unfortunate event occurred that attracted the interest of the public or the media.

These findings themselves did not necessarily mean that there was anything wrong with the organisation; the model could have been an inappropriate one, or the team could have misinterpreted the points made to them. It also has to be accepted that measures of performance for a system without a tangible end-product are difficult to define, and therefore difficult to measure; the output from this system could be expressed as *happier* or *cared for* people, abstract notions that defy precise quantification. However, the findings were supported by the issues that had been raised with regard to policy, poor communication and co-ordination, lack of resources to meet needs, and to some extent, the use of technology in support of the administrative functions, rather than related to the direct provision of services. The responsibility for resource allocation was split between two supportive Branches, and although arguably part of an operations system as a whole, these appeared to function in a semi-autonomous manner, albeit with limited communications with the Operations Branch on resource needs.

At this point in the study, the team were becoming slightly confused over whether or not the aims of the project could be met by this approach, or what the connection was between these organisational problems, and the potential for using technology. On reflection, the link can be easily made. Assuming the purpose of a study is to advise on technology applications, the first step is to employ the SSM to clarify where there are difficulties in achieving the objectives of the organisation, then to consider how these difficulties can be alleviated using the technology that is available. Taking the problem of measuring performance as an example, even in this sociological situation there are factors that could be regarded as indicators of success, such as the number of children taken into care, the timespan before clients return to normal unsupported life, or how this authority compares with others in the same peer group. The collection, analysis and dissemination of the necessary data could be achieved electronically, as part of a wider management information system. Similar conclusions could be reached about addressing other issues, particularly as there are few functions nowadays that technology cannot support in some manner. The approach also ensures that any improvements are seen in the wider context of the organisational goals, as demonstrated during the second phase of the Field Trial.

At the time, the connection was not clear, and the team were also concerned that many of the points raised during the interviews did not relate to the findings of the systems exercise, generally being minor problems or concerned with discrete areas only, such as the comments about the Home Help service noted on the rich picture. To resolve the dilemma, and to decide on the direction of the next phase, the obvious step was to consult the client about which changes he would consider *feasible or desirable*. The initial discussions indicated a reluctance to address major organisational problems at that time, and it was agreed to pursue only the less contentious matters that had been raised. This was a bit of a blow to the team, having developed grandiose ideas from the analysis; however, the early conclusions had a distinct effect on the final outcome of the study.

10.3.6 Feasible/Desirable Changes

Accordingly, a meeting was arranged with the Director and representatives from the Operations Branch, including Headquarters and Area office staff and front-line operatives. The meeting, which was also attended by Peter Checkland in a consultancy role, focussed on the technology implications of the problems that had been uncovered. The agenda included a presentation about the progress of the Field Trial, and an 'uncovering' of a rich picture summarising the main issues. A general debate was then held about the following:

1. The effect of decentralised policy on the needs for information at the SSD Headquarters, especially for monitoring and control purposes.

2. The volume of paperwork and administration associated with the delivery of care, and difficulties in the dissemination of information on procedures, policy, legislation, etc.

3. Concern over judging the total effects of changes in overlapping areas of social work, such as the effect on the Home Help service of changes in the policy for residential care.

4. Concern that the needs of front-line social workers for computer support had not been clearly identified, and that developments were 'management tools and for the elite'.

5. Difficulties in ensuring the proper dissemination of client information in Area Offices.

6. Concern over inadequate training and experience of front-line social workers.

7. Allocation, matching and accounting for resources, including financial, equipment, and manpower.

8. Communication and co-ordination problems within and between branches, and with external bodies.

9. Concern generally about the Home Help organisation.

Each of these problems were debated, and a simple weighting exercise carried out to gauge the perceived significance of the problems; everybody was asked to prioritise the issues on a scale of 1 to 10, with the totals indicating their relative importance. As the meeting was attended by staff from different levels in the Department's hierarchy, this had a flavour of democracy about it, although the final decisions were made by the senior people. However, there was a distinct polarisation of views depending on the part of the organisation in which the individuals were employed, and this exercise itself reinforced the need for better liaison between the front line and central functions. Eventually the discussion moved on to consider how technology could help; for example, whether a database of demographic information could be developed, allowing changes in the demand for services to be forecast, and the overall effect of policy changes estimated. The use of teletext systems for disseminating information, computer-supported training programmes for social workers, and many other potential applications were also explored.

The final selection was based on a process of elimination; the Director agreed that any further investigation should be targeted at front-line activities, removing from the list those problems that were mainly related to central functions. The remaining issues were then reconsidered, revealing that some of them were already being addressed. For example, a database for client records was under development which would allow them to be accessed via a network in and between Area Offices and the Headquarters; the Computer division of the County Council was devising a Home Help Organisers Allocation System, known as HOPS; a priority planning system, WPS, developed by Brunel University was being installed that would help in the process of identifying shortfalls in staffing resources, and also model the activities of Field Social Workers to give a clearer picture of their role in the community. The training question was one that the Department was fully aware of, and had recently injected some additional funds into their budget to address the problem universally.

It appeared from this that many of the front-line functions were already being considered for some computer support, with the notable exception of residential homes for the elderly, and day centres for training mentally handicapped adults. It was agreed at the meeting that the day centres would be a suitable arena for further study, more as a result of some hidden agenda that was not made clear to the team, than by a logical deduction process. Nonetheless, this was the area chosen by the client where some *action to improve* was considered worthwhile. It is worth reiterating at this point that the SSM does not aim to provide a scientific method for producing results, but essentially provides an agenda for debate about problems and solutions, and in this respect it had achieved its aim.

10.4 Application of the SSM - Phase II

The work that followed in the Adult Training Centres (ATCs) was a complete systems review in its own right. The team started with a new problem situation, structured it, developed root definitions and conceptual models, and came up with some recommendations for improving the situation using new technology. The exercise, taken with the findings of the first round of the project, was extremely revealing, and reinforced the value of the methodology for learning about an organisation as a whole before attempting to improve a small part of it.

10.4.1 Adult Training Centres - an Overview

Adult Training Centres provide day-time care and training facilities for mentally handicapped people, some of whom may attend for the majority of their adult life. Within the County there were a number of such Centres, providing a total of over a thousand client places, operated by County Council staff with assistance from outside voluntary agencies specialising in the care of the mentally handicapped. The resources of each Centre are controlled by a manager, aided by a deputy and instructional staff on a ratio of 1 instructor for 10 clients. Clerical and domestic assistance is also provided, and lunchtime meals are prepared by cooks, who also undertake some training duties. Clients are assessed on entry, and then at periodic intervals to determine their progress. The type of training provided depends on the degree of dependency of each individual, but is generally aimed at helping them to live in the community, and helping their parents or other relatives to care for them in their own homes.

10.4.2 Structuring the Problem Situation

To obtain a better understanding of the problem area, the team was advised to concentrate on one ATC, said to be representative of all those in the County, and the second round of fact-finding started. Mixing with the clients at this establishment was quite traumatic, and the team were thoroughly impressed by the work and the dedication of the staff involved in the training process. Mentally handicapped people have a permanent mental disability, sometimes accompanied by physical disablements; their disabilities inevitably result in behavioural problems, which in turn lead to problems in their relations with the community. They are able to attend a Centre from the age of eighteen, and some remain at the same place for thirty or forty years; compounding the problems of their handicap with problems of sustaining their interest, and that of the staff over what is often a whole lifetime..

Having settled in to this new environment, and completed a number of interviews, a rich picture and root definition were drafted by the team which seemed to reflect the main factors influencing the training of mentally handicapped adults. Although by this time more experience had been gained, the pictures were once again too rich, but nonetheless captured the essence of the chosen Training Centre.

Amongst the issues illustrated were ones relating to the client/ instructor ratio, felt to be too low; conflict with parents of the clients, who were often reluctant to give them the independence the Centre was encouraging; difficulty in developing and updating a training programme for each individual, particularly throughout their lifetime; timetabling difficulties brought about by the seemingly low ratio of staff to clients, making it difficult to provide cover in the event of absence. There were also problems over liaison with the Education Department of the Council on the provision of trained teachers in remedial skills, and on a more general note, it became obvious that staff were unsure about what they were trying to achieve, and knowing when they were achieving it.

10.4.3 Developing the ATC Model

The first root definition reflected the role that this training centre had adopted; that was, to provide somewhere for the clients to go during the day, where they could develop social skills and any latent abilities. The definition stated:

'A system owned by the Area Organiser, operated by instructors and support staff of Adult Training Centres, that occupies mentally handicapped adults during the daytime whilst developing and maintaining their potential ability and social acceptability in the community, within the constraints of County policy, statutory regulations and budgetary control'

A community viewpoint was taken, ie that the system aims to provide a place to go during the day, together with some training to draw out the potential of the clients, and make them more acceptable to the community. This seemed to be a rather negative statement of the purpose of the training system, almost a contradiction in terms, the emphasis being on occupying time and 'community protection', rather than providing training. However, it did reflect what the staff thought they should be doing, making it a relevant system to consider.

Some tentative conceptual modelling was carried out, which led to the conclusion that the definition was not an appropriate one for the training system as a whole, but simply reflected the role of one ATC. The team were uncomfortable with it, and the discomfort was caused by the reluctance of the staff at the centre to agree with it, and, more importantly, come up with any alternative. There seemed to be a lack of guidance from the Headquarters, and, although policy statements were examined, they were couched in such high-level terms that they were simply blanket statements of intent, for example:

"the Council is seeking to recognise the fundamental right of all such people to live independently in the community, determining for themselves their own future"

or

"mentally handicapped people have a right to be normal .... and are entitled to a home of their own"

These did not seem to provide any real direction for the manager of the training centre, nor did they help when trying to model a notional system that reflected their primary role. To obtain other views, it was agreed that a further two ATCs should be visited, each under the control of different Area Organisers. These had completely different styles of operation - one emphasised fund-raising activities (such as re-cycling waste) as a supplement to training, and the other was operating a small cottage industry, undertaking contract work for simple packaging jobs, or producing goods such as Christmas crackers. This is not intended to be a criticism of the Centres - faced with a vague statements about the policy for mentally handicapped adults, each manager had translated it in his own way under the guidance of the Area Office; the unfortunate consequence of this being that some clients were occupied more fruitfully than others. Incidentally, the team were most impressed with the cottage industry style operation of the third ATC. If normality was the aim, then this seemed the closest to achieving it, as the clients were employed for set hours, faced occasional peaks of work, produced goods to meet an actual consumer demand, and were given a weekly pay packet, generally being treated the same as people regarded as 'normal'.

Once again there were some misgivings about how this related to the use of technology and to the objectives of the Field Trial. However, in terms of testing the method this turned out to be a very revealing and worthwhile exercise. It seemed, while there was such a diversity, it would be difficult to develop computer systems that had a general application to the training of mentally handicapped adults, except in relatively small areas, such as common returns made to the Headquarters. At the local level, even the bookkeeping requirements were different, so the systems thinking was continued to see if it would help in the development of ideas about the needs of the Centres as a whole.

In the course of the subsequent examination some of the earlier modelling exercises were reviewed, and it was reassuring to find that these had revealed the potential root causes for the problems that were now being revealed. The findings with regard to coordinating and policy interpreting, and the lack of critical success factors, were all relevant to the situation currently being examined. The root of the problem seemed to be at the top of the tree! Without a system to translate policy statements into statements of the activities needed, and indeed the resources required, effective co-ordination could not be achieved, as there would be inadequate information to give proper direction to the front-line services. Without clear directives and some idea of how to measure success, then each of these front-line systems would *do its own thing*, with varying degrees of success. Given that there did not appear to be a coordinating system anyway, there could be little hope of achieving a consistent level of service.

10.4.4 The ATC Conceptual Model

To return to the development of a relevant system for training mentally handicapped adults, it was agreed that the excursion into organisational problems, though revealing and intellectually satisfying, did not help with finding ways of making improvements. It wasn't however a wasted distraction, as it indicated that, for any changes to be cost effective, they would have to be developed in a way that didn't reflect the activities of just one ATC. Accordingly, a new W was taken, this time from the point of view of an individual client, or possibly a concerned person acting on behalf of the client. This resulted in a notional system to *identify and satisfy the needs of mentally handicapped adults*, which, being independent of the activities observed at any of the Centres, achieved the desired neutrality. The concept of normality was introduced to recognise the intent of policy statements, although it was still unclear how this idea could be defined or measured. The definition included the following:

'A system owned by ....etc that identifies and provides the help that a mentally handicapped adult needs to attain and sustain a way of life that equates to that led by persons without mental impairment ....etc'

The aim was to define a system to treat each client individually, by identifying his or her training needs against a yardstick of normality, and providing a programme to meet and sustain those needs. In developing the conceptual model, it was also accepted that, for many clients, there would never be a time when normality was reached, consequently the process would be continually iterative; having made a few steps forward, the client re-entered the system and was re-assessed. The model had two main elements, one for *identifying* training needs, and the other for actually providing the training, or *satisfying* these needs.

10.4.5 Examining the Situation

The final part of the exercise was to compare the model with the real situation, drawing on the knowledge gained during the visits to the ATCs, and taking account of the requirement to identify where information technology could be used beneficially. To do this, all the sub-systems from the model were listed, and each of the lower-order activities identified, such as assessing ability, finding out about the wishes of parents, assessing community related needs, producing a client programme, and monitoring progress.

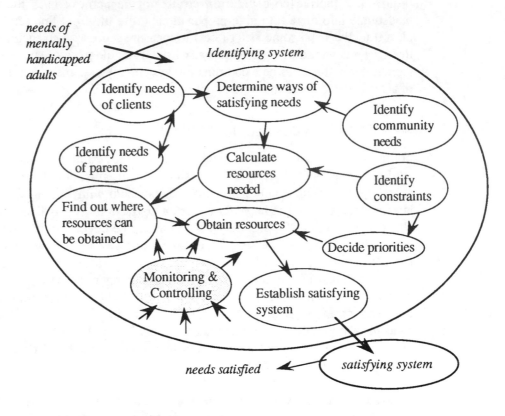

Fig 10.6 - Extract from ATC System Model

The training sub-system model had activities concerned with knowing how to train, assessing the resources needed, knowing where to get them, getting them, prioritising, and producing a training programme for the centre as a whole. An extract of the model is shown in Fig 10.6.

A Checkland-style matrix was then constructed so that the existence of each activity and its effectiveness could be considered. Certain activities were clearly reflected in the real situation; for example, on entry to an ATC each individual was assessed using a detailed list of ordinary things that a normal person does, such as cleaning teeth, washing, dressing, catching a bus etc. At periodic intervals this was up-dated to determine how much progress had been made. However, because of the level of detail, the numbers of clients and long duration of their stay at the Centre, it was difficult to produce and update a personalised programme for each client, or to bring together all similar requirements to produce a comprehensive training programme for clients with similar needs. It was concluded that, although many of the required activities existed, either formally or informally, most could be improved in some manner by the use of new technology. To clarify the types of information each activity needed, and to provide a basis for checking how available software could be used by the ATCs, each element of the model was expanded until the information requirements could be listed, as summarised in Fig 10.7. (This approach was the basis for the information systems enquiry method detailed in Chapter 11.)

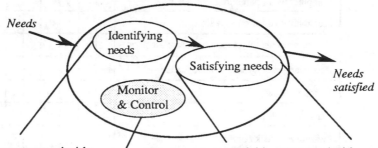

Activities concerned with:

Assessing ability
Assessing wishes of parents
Producing client programme
Monitoring progress

Requirements for information:

about clients
about parents
about resources
about training (national and local knowledge)
about constraints
about success factors (normality ?)

Activities concerned with:

Knowing how needs can be satisfied
Assessing resources
Knowing where to obtain resources
Obtaining resources
Knowing constraints
Prioritising
Producing training programme

Fig 10.7 - Expansion of ATC Model

10.4.6 Options for Improvements to ATCs

The analysis resulted in the conclusion that the primary activities of the ATCs, ie those that identify and satisfy the needs of clients, and other secondary activities, were found to exist in each of the ATCs that had been visited. However, as individual managers had different priorities and made different use of available resources, the training received in some centres could be more appropriate than in others. The main weakness of the system as a whole (ie Countywide), therefore, was the lack of common objectives and goals which would guide the managers in fulfilling their primary role. This fundamental weakness, which was outside the control of individual managers and Area Organisers, reduced the potential for cost-effective IT applications, as each ATC would have to be considered separately. Development costs would be greater, and additional advantages, such as improved communications and the use of standard procedures, would be difficult to achieve.

APPLICATION	1	2	3	4	5	6	7	8	9	10	11	12	Total
Recording and analysing assessment results				x		x		x		x		x	4
Computer file of client details			x		x								9
Computer file of parent/relative details			x										7
Electronic mail for cross-pollination			x										3
Match assessment with resources, produce prgm.		x		x	x		x						15
Accounting packages for holiday funds						x		x					2
Wordprocessing	x					x		x					3
Class timetables		x				x			x				2
Inventory & stock control	x			x		x			x				2

Total calculated by multiplying 'x' by Relative Value Weighting,
ie 4 points for primary function, 3 for secondary, 2 for direct
support and 1 for indirect support.

Fig 10.8 - Application/Function Matrix

Nonetheless, by assessing potential computer applications alongside the system model it was possible to develop ideas about how technology could be used to support the primary objectives of the ATC organisation. Following an investigation of the software market, a further matrix was used to weight each available application in relation to the functions they could support. An extract from this matrix is shown in Fig 10.8.

This indicated that there were three possible areas where technological improvements could be effective, summarised as:

- *Strategic,* or those that in the long term would enable significant improvements in client care to be achieved, such as the production of customised client programmes.

- *Operational,* or those that would directly aid the current methods of providing training for clients, such as educational software, training packages for the development of motor skills and co-ordination.

- *Support*, or those that would assist the support functions of the training centres, such as word processors, and spreadsheets for accounts.

The benefits of each group of applications, and their possible costs, were also assessed. Maximum benefits could be anticipated from the strategic use of IT. However, without a common ATC policy, this would not be sensible or cost-effective. The operational applications would be effective and provide direct benefits to clients, but, unless used within the framework of a coordinated client training policy, then the benefits overall would be reduced. In the support service areas, generally the activity rate (in relation to information usage and change) was too low for any particular activity to justify computer support, but taken overall some advantages could be identified in terms of reducing the administrative load of the Centres. Once again, individual applications to suit the needs of each Centre would have to be considered, at a greater cost than introducing applications for general use.

To complete the exercise, the County computer strategy was examined so that some practical advice could be given to the Director of Social Services on potential purchases. As a result, it was possible to recommend the use of IBM microcomputers to support the strategic applications, which could also be used for the support functions, and, as the IBM was the County standard micro, could form part of any network that might be developed in the future. The BBC micro was recommended for the Operational use because of the wide range of educational software that was available, and also because support for the application could be obtained from the Education Department of the County Council.

The final report included a detailed breakdown of the actual application packages, and also expressed concern about the lack of a coordinated approach to the training of mentally handicapped adults, which would mean the development of 'customised' packages for each Centre. Advice was prepared on a strategy for implementation, which would recognise, amongst other things, that the extent of computer literacy in the Centres varied considerably, depending more on the interest of individual staff than any common policy for training.

Using the hypothesis that the owners of these front-line systems were the Area Organisers, and therefore it was they who needed convincing before the Director would consider taking action, the next step was to have a full and frank discussion with the Area Organiser who had asked for the examination of the ATCs. Whilst accepting the logic of the argument, he was unhappy about the choice of equipment, having a particular desire to introduce a wordprocessor that he was familiar with, but which did not feature on the County's approved list of microcomputers.

After further discussion, a compromise was reached and both types of equipment were introduced on a pilot basis. Eventually a number of ATCs were also equipped with BBC micros for use in training; and a policy for the development of customised programmes for all mentally handicapped clients agreed. On the broader front, the Department has since been re-structured and the number of Areas effectively reduced, at the same time moving some of Headquarters functions closer to the front-line. There is also a growing awareness of the need to set and to measure standards of performance throughout , mainly as a result of recent initiatives on the provision of care in the community.

10.5 Conclusions

As far as the Field Trial was concerned, both of the objectives had been achieved, ie the SSM had been tested, and the client advised on possible uses of new technology. The methodology had proved its worth in a number of ways, the analysis of the Operations Branch as a whole alerting the team to the general lack of consistency in service delivery and standards, encouraging ideas that were not influenced by the style and operating practices of any one establishment, and the development of models that reflected the fundamental purpose of the training enterprise.

In effect, the original question being addressed had changed, subtly but significantly; instead of *'where can technology be used'*, the analysts had been encouraged to ask *'how can technology contribute to the accomplishment of the organisation's objectives ?'* Furthermore, the knowledge gained from applying the SSM ensured that the context of the question was clearer. The combination of a better question and better informed analysts had increased the chances of arriving at solutions suited to the needs of the organisation and the people involved.

Further conclusions were drawn from the exercise that were relevant to the total FAOR package, and to the value of the SSM for focussing a study in the early stages. The initial examination of the Operations Branch had revealed a variety of problems, none of which seemed particularly pertinent to the question of computer usage. At the time it seemed that a sharper technique was required to achieve this focussing, possibly by considering the level of information dependency of each functional group, as indicated by such factors as the volume of letters received or produced, telephone calls, files held and/or used, and so on. This approach had been tried on an earlier occasion, and, although reasonably effective was extremely labour-intensive, involving a large number of people counting pieces of paper passing between various groups. On the basis that 'who uses most gets priority', focussing could be achieved in this way. However, with an organisation as large as the one addressed during the Field Trial, paper-counting would probably still be going on.

The SSM does not provide a precision tool to direct a study at technology improvements, but, being primarily concerned with problem-identification, guides the analysis towards those parts of the organisation where there is a need for an improvement of some sort. A decision can then be made about which problem to address, and *how* to effect a solution, in consultation with the client as part of the stage 6 debate about *feasible/desirable* changes.

Once this decision is taken the SSM can assist with a 'top-down' analysis of the selected part of the organisation, in a manner similar to that used during the ATC phase of the Field Trial, working from an agreed statement of the primary role and expanding the models until the logical information requirements become apparent, as explained in Chapter 11.

As a final footnote to this Chapter, since the time of the Field Trial there have been significant changes that affect the provision of the social services. Government pressure on all local authorities to become more cost-effective, and major initiatives to improve community care, are just two factors that have caused a major disturbance to the social services system as a whole. More establishments, whilst remaining within the control of the Council, will become virtually autonomous, and others will be taken over by the private sector. The needs of individuals will be examined more thoroughly when determining the type of support required, and personal care programmes will become the norm. The Social Services Department will gradually assume more of an enabling role, rather than that of a main care provider, with additional responsibilities for inspecting all establishments and service functions within its geographical boundaries. The models developed during the study now seem even more pertinent; changes of this order suggest that the need for good communications, clear policy statements, and sound monitoring and control mechanisms will be even greater in the future.

11 A Method for Information Systems Enquiry

11.1 Introduction

When embarking on a study that addresses the information needs of an organisation it is often difficult for the project leader to ensure that the project is both cost-effective and comprehensive. Large or small organisations may comprise a variety of functional groups, some undertaking essentially manual tasks (eg assembling or manufacturing components), others involved with sales, marketing, costing etc, and others with indirect support work such as typing, filing, and the distribution of correspondence. Each function requires information to operate, and produces information for others to use. The information base is diverse and complex, and, to the outsider, can be difficult to understand and difficult to investigate in any structured or systematic way.

As a consequence, there is either a tendency to overcompensate and carry out too much fact-finding in a rather loose fashion, or to simply focus on specific information processes with a view to introducing new technology. In neither case does a clear picture emerge of the true value of any improvements in relation to the the organisation as a whole, and often the analyst is left with an uncomfortable feeling that more could have been achieved if the situation had been approached more methodically.

The Method for Information Systems Enquiry, known colloquially as MINSE by the research project team, specifically addresses the problem of structuring the approach to information studies. It uses systems thinking to develop ideas about *what* information is needed to achieve a defined purpose, ideas that are independent in the first instance of *how* this information appears in practice. In the process an information model is built up on a computer database which is then used as a framework to explore the situation and identify problems at a number of levels (Fig 11.1). At the outset, it also considers issues within the organisation that may indicate areas of potential improvement. MINSE has been developed using unsophisticated database software, software that is generally affordable and available to most practitioners, and does not require any depth of knowledge or training. In the sense that it is an enquiry method, this is considered sufficient; extending the analysis to produce a comprehensive information model of an organisation is possible, but this would require the use of additional software and then becomes a more specialised operation. However, the principles are the same regardless of the extent that MINSE is applied, and only the purpose of the investigation will determine the depth of analysis needed and the amount of resources that are worth employing.

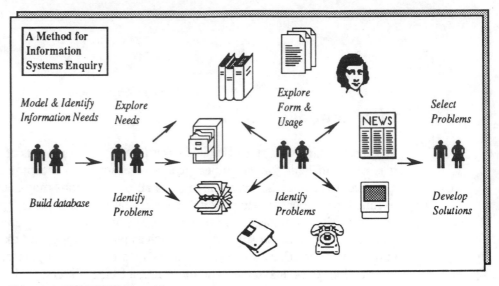

Fig 11.1 - The MINSE Approach

This chapter describes the research project that was carried out to develop the method. Like the FAOR Field Trial (Chapt 10) it took place in an establishment that was under the control of a Social Services Department. This establishment, a residential home for the elderly, was chosen deliberately because of its multi-functional nature, operating in effect as a self-contained enterprise, responsible ultimately to the parent department but encouraged to manage its own affairs on a day-to-day basis. In addition to fulfilling its primary function (ie providing support for elderly persons in a residential setting), the manager is also concerned with all the financial aspects of the home; with equipment purchases, catering, maintenance of premises and gardens; deploying staff, accounting for their time and arranging appropriate payments; resolving personnel matters, and so on. In other words, the overall management of an organisation that is in many ways similar to a small hotel or guest house, where most of the guests are extremely frail elderly people. This multifunctional environment was considered suitable for developing a method that would be equally applicable to a wide range of organisations, albeit with totally different business objectives.

11.1.1 Overview of the Method

MINSE was developed from the hypothesis that ideas about information needs can be derived from an agreed systems model that reflects the primary purpose of an organisation, expressed as the transformation in the root definition of the system. The information needs would be identified by taking the model through various stages of decomposition until such time as it is possible to define the information requirements of each activity that contributes to the transformation. Once the *systemic* information base has been clarified, this would be used as a template to explore, and where necessary improve, aspects of the *actual* information base of the organisation.

MINSE makes full use of the SSM stages as summarised in Fig 11.2. A series of interviews is carried out to clarify relevant factors about the organisation, which are then analysed and illustrated in a rich picture. After discussions with the client, a neutral viewpoint is selected as the basis for a root definition and conceptual model. The model is progressively decomposed until the lowest order of activities can be identified, at which point the information needs of each activity are listed.

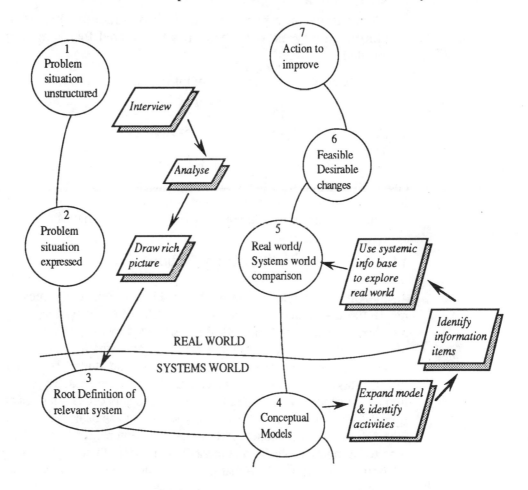

Fig 11.2 - MINSE and the SSM

A standard database application (eg SMART on the IBM PC, WORKS or JAZZ on the Macintosh etc) is then used to record and code each information *need* by type (eg Financial Information - FI, Personnel Information - PI, and so on). Further codes are added to indicate the functional groups that use the information, the source and/or destination of the information item, and whether it is generated from within the system, or received from an external source.

The analysis reveals the information that could or should exist if the model is a fair reflection of the real world activity. Using the database sorting facilities to regroup the information as required, it is then possible to explore the actual information base of the organisation in a number of ways. For example, to:

a. Examine a particular category of information where there appears to be a problem.

b. Concentrate on a particular activity or function that uses a variety of information types, but where it is considered that some improvement is needed.

c. Investigate the communication of information between all or selected functions.

11.2 The Research Project

A project to research the method was undertaken over a period of approximately one year on a part-time basis; at various stages a number of different practitioners from management services disciplines were involved, each receiving some informal training on the principles and practice of soft systems analysis.

11.2.1 Overview of the Organisation

In the United Kingdom, local authority Elderly Persons Homes (EPHs) provide residential accommodation for elderly people who can no longer manage in their own homes. There are similar facilities provided by the private sector, but, in terms of the number of residents, these are generally smaller than the local authority Homes where the average size at the time of the study was around 48 beds. In recent years there has been a tremendous drive to keep elderly persons in their own homes for as long as possible, by giving close support to relatives and friends and by providing support from community services such as home helps and 'meals-on-wheels'. As a consequence, elderly persons entering residential accommodation tend to be more physically and mentally dependent than in the past, and the type of support needed is similar to that given in a nursing home, rather than a guest house or private hotel.

Working in an EPH, either as a member of staff or as a consultant analyst, can be a moving and traumatic experience, and is far removed from more typical analysis work in office and business situations. Staff are dedicated to their work and generally regard it as vocational, having to cope with frail and sometimes difficult people in their declining years, and, on occasions, the reality of death. In these circumstances the analyst cannot help becoming involved, particularly if he or she has elderly relatives in similar situations, and it is sometimes difficult to retain objectivity. It is also necessary to recognise the fact that the organisation is the *home* of the residents, and their dignity and privacy must be respected and given priority over analytical considerations whenever necessary.

The nature of the organisation and the staff it employs tends to lead to a greater commitment to any investigation and subsequent changes, particularly if it is felt that they could eventually benefit the residents. In this type of situation, once a study is underway and the initial reservations are overcome, staff tend to talk freely to the analysts and there is a high level of client participation in all stages of the exercise.

11.2.2 The Organisation Structure

By way of background information, some EPHs operate with a hierarchical structure of an Officer-in-charge (ie the manager) with 3 assistant officers, one of whom would be appointed as the Senior Assistant and act as deputy for the Officer-in-charge when required. Care Assistants are employed to provide direct support to the residents with such matters as toileting, dressing, washing and so on, supervised by a Senior Care Assistant. Cleaning and other domestic duties are carried out by Domestic Assistants, and every EPH has cooking staff, clerical support, and help with gardening and building maintenance, generally provided by a gardener/handyman. The structure is summarised in Fig 11.3.

The EPH which featured in the MINSE project provides residential facilities for more than sixty elderly people, some in advanced stages of dependency, and it is in full-time operation throughout the year. It also offers a limited number of places for the elderly in the community to attend as day clients. The Officer-in-charge (OIC) is under the line control of the local Social Work Area Office, but is semi-autonomous with regard to the day to day running of the EPH.

The type of work undertaken in the EPH and the functional groups responsible can be summarised as follows:

- Management and personnel matters - OIC and Assistant Officers.

- General administration and clerical work - officers and clerical assistant.

- Care of the residents and day clients - officers and care assistants.

- Cleanliness of accommodation and offices - domestic assistants (supervised by officers/Senior Care Assistant).

- Catering and food preparation - cooks (supervised by officers).

- Routine maintenance of buildings and gardens - gardener/handyman.

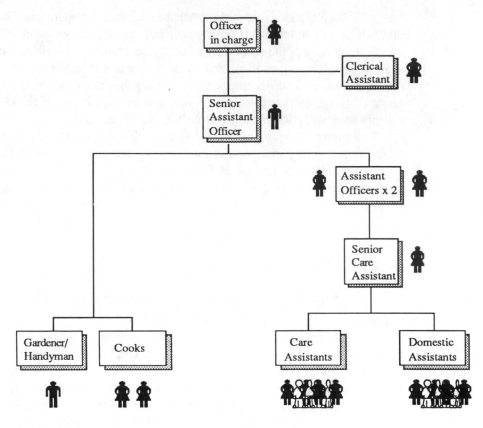

Fig 11.3 - Example of an EPH Organisation Structure

11.3 Stage 1 - Exploring the Situation

The first stage of the process involves a conventional fact-finding exercise, where representative staff are interviewed to determine views about the purpose of the organisation, the role of the staff, and any problems that are pertinent to the study. The interviews are then analysed, using the database analysis technique described in Chapter 5, and a summary of the situation prepared as a rich picture. This is used to develop ideas about relevant systems and construct a neutral conceptual model, which is then expanded to identify activities and related information.

11.3.1 Information Collection and Analysis

At the outset of the research project, after initial discussions with the Officer-in-charge, a formal briefing was given to all staff explaining the aims of the project and the approach to be taken, giving the opportunity for any questions to be raised. The reaction at the meeting was rather subdued, but at the subsequent interviews most staff expressed their views freely about the way the EPH operated and the purpose that it served. It was noticeable that many of them made emotive comments when addressing this question (for example, *"a last resort"*, or *"a place for old people to go when nobody else wants them"*), a reflection possibly of their personal involvement with the residents of the home and concern for their well-being. The interview analysis, a process that invariably requires a great deal of judgment on the part of the analyst, was even more difficult as a result, particularly when attempting to derive an uncontentious root definition at a later stage.

Nonetheless, the examination helped to focus attention on relevant facts and issues about the situation, although drawing the rich picture was quite difficult, and there was the usual tendency to include too much detail. An extract from the picture is shown in Fig 11.4.

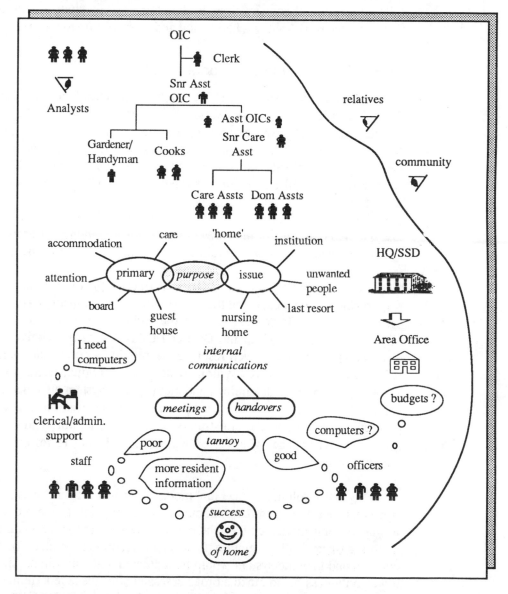

Fig 11.4 - Extract From EPH Rich Picture

The early pictures were drawn on a flip chart with the help of staff from the home, and later reconstructed using computer graphics. The picture highlighted amongst other things the perceived role of the EPH, differences of opinion between members of staff about the value of existing information and the manner of communication, and the general consensus of opinion that the operation was a success if residents are happy (shown as 'smiling faces' in the illustration). The picture also served as a useful summary or overview of the whole EPH situation, reflecting its position in the surrounding community, liaison links with social workers and relatives of the clients, and its relationship with the area office and the department headquarters. The views of residents were also taken into account, following informal conversations in the course of the study with those who were able to understand what was happening and were willing to voice an opinion.

11.3.2 Constructing the Model

As in most studies using the SSM, the development of the root definition was a difficult and lengthy process. Although it is easy to observe what is going on in practice, constructing an explicit statement of what the system *is* that is being considered inevitably gives rise to problems of interpretation and semantics. The nature of this particular situation added an emotional element that made the task even more difficult, and many subjective statements were made about the transformation, statements that appeared to reflect the concern of the analysts for the recipients of the service (for instance, *making old people comfortable in their last few years*, and *providing caring services for terminally old people*). Eventually, after prolonged discussion, the following was agreed as a working definition:

'A system owned by the Area Organiser (O) - (ie the Social Services officer responsible for the area, using the *ownership hierarchy* theory explained in Chapter 5), operated by officers and staff of the EPH, other designated Social Services officers, associated staff of the District Health Authority, relatives and friends of clients (A), which uses given resources to provide a congenial dwelling place for referred elderly people (C), within which the individual needs for physical and emotional support are determined and met (T). The system also identifies and meets the needs of elderly people in temporary attendance as day clients (C), and is continually subject to the constraints of the local authority and other statutory regulations (E).' The W assumed was a neutral one, ie reflecting a public or common viewpoint .

The first-resolution conceptual model shown in Fig 11.5 was then constructed from this root definition. Consideration was then given to the requirements of the formal systems model, eg the relationships between the system components (taken in this case to be information flows), the resources needed for the system to operate, and how the system would be regulated and controlled. At this high level, it was only possible to make a broad assessment of these requirements, and further ideas were developed as the study progressed to lower levels of detail.

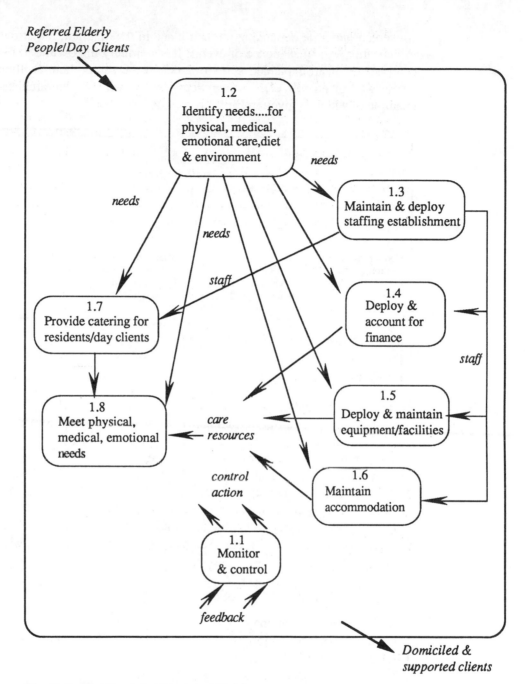

Referred Elderly People/Day Clients

1.2
Identify needs....for physical, medical, emotional care,diet & environment

needs

1.3
Maintain & deploy staffing establishment

needs

needs

staff

1.7
Provide catering for residents/day clients

1.4
Deploy & account for finance

staff

1.8
Meet physical, medical, emotional needs

care resources

1.5
Deploy & maintain equipment/facilities

control action

1.6
Maintain accommodation

1.1
Monitor & control

feedback

Domiciled & supported clients

Fig 11.5 - First Resolution Conceptual Model

Each sub-system of the model was given a decimal code so that an audit trail could be established for the subsequent detailed analysis. As a matter of procedure, it was agreed that all the monitoring and control components of the model would be given the suffix '.1', whereas all other components would be numbered in a convenient order, ie depending on their place in the diagram, rather than annotated to indicate a logical sequence of events. It is normally impossible to display all the components of an expanded system model in the same plane (without using the whole wall of an office!), so they were also shown as a hierarchical scalar, an example of which is given in Fig 11.6.

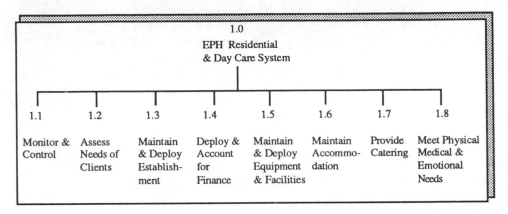

Fig 11.6 - Scalar Summary of Conceptual Model

11.3.3 Expanding the Model

To progress the analysis to the level at which activities could be identified, a root definition was developed for each sub-system in turn, and a second-level conceptual model constructed. Taking sub-system 1.7 for example, the lower order systems needed to *provide full board catering for residents, and occasional meals for day-clients* were considered, and the second-level model shown in Fig 11.7 was constructed. Each component of these models was then expanded in the same manner to identify the associated activities. To avoid losing control of the analysis process it was necessary to list the activities in tabular form as well as show them diagrammatically (Fig 11.7). For example, the activities associated with sub-system 1.7 were listed as:

1.7.1	Monitor and Control
1.7.1.1	Ensure hygiene
1.7.1.2	Ensure nutrition
1.7.1.3	Obtain feedback
1.7.1.4	Take control action
1.7.2	Obtaining (ingredients)
1.7.2.1	Prepare menus
1.7.2.2	Determine ingredients/supplies needed
1.7.2.3	Determine source of supplies
1.7.2.4	Obtain ingredients

1.7.3	Delivering (meals)
1.7.3.1	Prepare food
1.7.3.2	Cook and serve
1.7.3.3	Clean up

In the case of the MINSE project, where the intention was to record activities on a computer database, this also served to validate the codes given to each item before entering them into the computer.

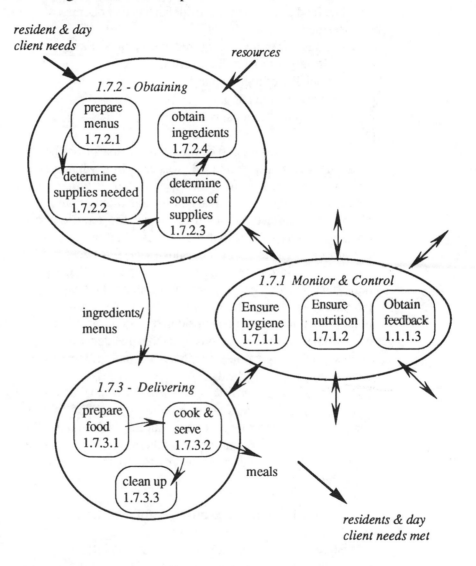

Fig 11.7 - Second Level Model of Catering Sub-system

11.3.4 Identifying Related Information

Having identified the activities associated with the sub-systems, it was then possible to start considering what information was relevant to the performance of each activity. To illustrate this, consider what information is needed by, or produced from, the activity *preparing menus* (1.7.2.1), such as:

The number of residents requiring meals each day
The number of day clients taking meals on a day-by-day basis
The food preferences of residents and day clients
Residents' dietary needs
The number of meals required at each meal session
Budgetary provision for catering
Availability of ingredients
Sources of food supplies
Cooking instructions, etc

The most difficult part of the exercise, apart from handling the sheer volume of data that was being produced, was deciding what was meant by 'monitor and control' for each system element, and then drawing out the related information. There is a tendency for analysts to add monitor and control to system models almost as an afterthought, just to meet the requirements of the formal model. Experience has shown that many real situations are lacking in mechanisms for monitoring what is going on, and taking the necessary control action when the required standards aren't being met. Ergo, developing ideas about measures of performance, feedback and control mechanisms is arguably the most important and potentially productive part of the exercise. As a rule of thumb, this element in the main system model can be considered as monitoring the performance of the system as a whole (for example, ensuring that residents are happily domiciled and day clients receive appropriate support), and setting, then maintaining, the standards of each sub-system (Fig 11.8).

This idea of a *hierarchy of control* allows the analyst to focus on each element in turn, and derive ideas about monitoring and controlling at various levels. With regard to the performance of the catering sub-system (1.7.1), this indicated the requirement for information such as the following before adequate control could be achieved, together with mechanisms for feedback on the extent that standards are complied with:

Required standards and practices for hygiene and safety in the kitchen.
Cleaning standards for the kitchen.
Standards and practices for freezing and defrosting food.
Information on balanced meals for the elderly.
Nutritional standards.
Spending constraints, ie budgetary control information.

It is also worth noting that, as the analysts became immersed in the detailed examination, there was a tendency to forget that this was still part of the *systems thinking stage*, and that the models, activities and related information needs were those that were desirable in *systems* terms, and not did not necessarily reflect the real situation. The conceptual line between the systems world and the real world was crossed at a later stage when the information model was used to explore what was actually happening in practice.

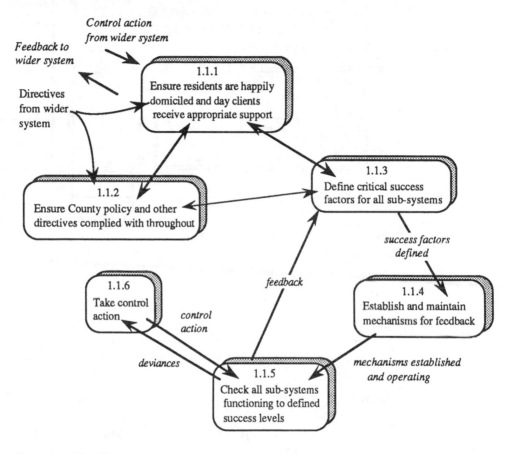

Fig 11.8 - EPH Control System

11.4 Using the Database

The idea of using standard (ie flat-file) database software arose from the earlier experiments with interview analysis. It became obvious as the system model was expanded to activity and then information level that the volume of data produced was becoming impossible to handle and comprehend using manual methods only. An impression of the extent of expansion and subsequent volume of data can be gathered from the illustration in Fig 11.9, which also summarises the stages of the exercise up to this point.

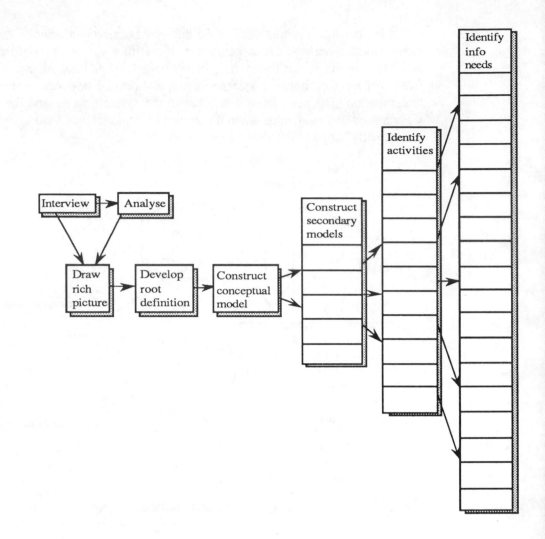

Fig 11.9 - Summary of MINSE Expansion

The software utilised for the interview analysis (ie JAZZ) operated quite successfully on the semi-portable (luggable?) Macintosh Plus, which could be taken on site to encourage the full participation of the client group. In the absence of more sophisticated software, and taking account of the desirability of direct client involvement, it was decided to try and make use of the same programme for record-keeping and analysis. (Having the Macintosh on site also helped to establish a good rapport with the EPH staff, who occasionally commandeered it to help with administrative work, rewarding the project team with small tokens of appreciation, such as cups of tea and bacon sandwiches !)

In a similar manner to the interview analysis technique, a record was set up on the database for each activity using the numerical codes as an identifier, and a second field used to type in all information needed or produced by that activity. These were first written down in draft form, then refined on input to the database.

An example of the database is shown in Fig 11.10. As with most research projects covering unfamiliar ground, it was not clear at the time how this would help the subsequent investigation, but it met the immediate need to record the results in a concise and tidy manner. The next stage therefore was to consider how this large systemic information base could help in exploring and understanding what was happening in practice, and how to link the system activities with the functional groupings that existed in the EPH .

Activity	Sub Code	Related Information
Client selection	1.2.2.1	Name and address of Doctor
	1.2.2.1	Name of client
	1.2.2.2	Personal history of resident
	1.2.2.2	Medical history of resident
Client assessment	1.2.3.1	Interests and hobbies
	1.2.3.1	Residents views and preferences
	1.2.3.1	Other residents names

Fig 11.10 - Example of MINSE Database

11.4.1 Coding by Type

After detailed examination of the information fields, it was felt that it would be possible to categorise them by type as a basis for selecting areas for further study. For example, many of the system components made use of, or generated, information about residents - their names, their preferences, their current state of health, and so on. In the real situation, assuming that the model was a reasonable reflection of the real-world activities, this information would be used by a variety of functional groups, and an investigation without a master checklist would be complex and potentially inaccurate. Grouping all resident-related information would provide this checklist for probing the availability, quality, and communication of the actual information about residents in the EPH. Accordingly, all the records in the information field were scrutinised, and coded to one of the following categories:

RI   Resident Information
FI   Financial Information
PI   Personnel Information
CI   Catering Information
AI   Accommodation Information
EI   Equipment Information

Certain information within these main categories was considered to be primarily concerned with monitoring and control activities, and, where this could be specifically identified at this stage, it was coded accordingly (ie COI). The EPH information types are summarised in Fig 11.11.

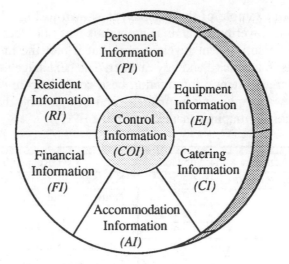

Fig 11.11 - EPH Information Types

11.4.2 Coding by Functional Group

To make the link between the activities, information and functional groups, further fields were added so that the possible users of the information could be identified. This effectively established a matrix on the database (Fig 11.10) so that an annotation could be made to indicate whether each item of information, *in relation to the activity described,* was relevant to the functional groups that existed in the EPH, ie the Officers (OICs), Clerical Assistant (Clerk), Care Assistants (CA), Domestic Assistants (DA), the Cook, and the Gardener/handyman (Gdnr). The annotations 'x' and 'o' (relevant or not relevant respectively) provided the basis for an alphabetical sort within each functional category.

Sub-system codes   *Category codes*   *Functional groups*

Activity	Sub	Related Information	Code	OICs	Clerk	CA	DA	Cook	Gdnr
Obtain food	1.7.2.1	Numbers of residents	RI	x	x	o	o	x	o
	1.7.2.1	Residents food preferences	RI	x	x	o	o	x	o
	1.7.2.1	Budgetary provision	FI	x	x	o	o	x	o
Provide meals	1.7.3.2	Kitchen staff availability	PI	x	x	o	o	x	o
	1.7.3.2	Cooking times and methods	CI	x	o	o	o	x	o

Fig 11.12 - Illustration of MINSE Database with Functional Codes

11.4.3 Coding by Source/Destination

The most difficult part of the process was attempting to identify information links between the system components, ie the source of information required by the activity, and the destination of the information produced. The use of Net Theory, often called 'Petri nets' after its initiator C A Petri (*Introduction to Net Theory* - Brauer 1980) was considered, but discarded as being too specialised to meet one of the research project aims, ie to produce an enquiry method that could be used by practitioners with the minimum of training. Consequently, simple codes were devised and entered in separate fields to indicate, in broad terms, the source or destination of information items.

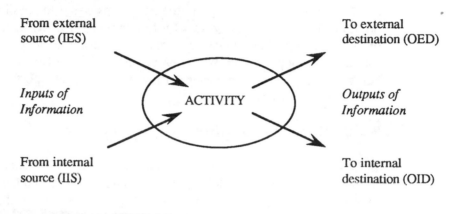

From external
source (IES)

To external
destination (OED)

*Inputs of
Information*

ACTIVITY

*Outputs of
Information*

From internal
source (IIS)

To internal
destination (OID)

Fig 11.13 - Source/Destination Codes

The codes used are summarised in Fig 11.13; *internal* in this instance referred to any activity within the EPH system model, and *external* to those of the wider system of interest or surrounding environment, eg the Department headquarters, Area Offices etc. At this point, the limitations of the database were realised, ie whenever information was received from, or transferred to, more than one system component, it was necessary to add one additional record to the database for each source/destination. This in turn made it necessary to repeat the entry for all related activities, causing a great deal of duplication of records. From this it was concluded that, although the principles of identifying systemic information links by this method appeared sound and it could be achieved to a limited extent by using the facilities of a flat-file database, it would not be cost-effective to pursue without access to more sophisticated technology.

11.4.4 Sorting Options

To summarise, at this point in the analysis the database had been set up listing all activities considered systemically desirable to achieve the transformation defined by the root definition, identified by description and by code. Against each activity the information required for successful operation, or produced during the activity, had been listed. To this database had been added further codes to indicate the type or category of information, the functional groups in the EPH using it, and its source or destination. This exercise established the potential for sorting the database in a variety of ways, for example:

- by activity
- by category
- by functional group
- by source or destination (subject to the limitations explained earlier)

Further sorting within these primary groups was also possible, for instance to bring together specific *categories* of information (eg FI, RI) within the ordered list of *activities*, and so on. It was later found to be useful to set up separate databases for each functional group, which prevented earlier versions of the full database being corrupted as experiments in sorting and producing reports were carried out.

11.5 Data Validation

Before moving on to make use of the database, it was necessary to assess whether or not the data itself was valid for the organisation being studied, bearing in mind that so far the exercise had been a systems-thinking one, rather than a study of what was happening in practice. The validation was carried out by first producing in report format a list of activities and associated information that appeared relevant to each functional group, then asking questions of selected staff on the basis of this list. As a result, it was possible to draw certain conclusions:

1. The systemic information base was of a general nature, and did not indicate the form or usage of the *actual* information objects in the EPH. After some discussion it was realised that this was a *what/how* problem, ie the analysis had revealed *what* information was needed, but not *how* it appeared in practice (see also page 188). Although this meant that the investigation required a further step to make the link with what was happening in practice, it also implied that the systemic information base would remain valid even if changes took place to the form of the actual information items.

2. Some systemically-desirable information, although part of the wider system, was not relevant to the actual activities carried out in the EPH. For example, the analysis indicated that domestic staff should be deployed on the basis of information about the cleaning areas of rooms in the establishment, but deployment was *actually* based on the experience of the officers, tempered by such considerations as staff availability and custom and practice. However, the domestic staffing levels overall had originally been calculated using room areas as an indicator of workload, confirming the validity of this information in terms of the wider system, although not related to the EPH activities.

3. There were noticeable omissions in some areas, primarily those concerned with administrative tasks. This was considered to be due to the limited involvement of the officers and the clerical assistant in certain parts of the analysis. The MINSE approach requires a detailed investigation which moves beyond the intuition of the analyst, and can therefore only be accurate with a continual input of local expertise which was obviously missing during some of the phases of the research project.

4. In many respects, the level of detail appeared too fine for an organisation where most internal communication was of an informal nature, and unlikely to be suitable for enhancement by electronic means. Conversely, where there is potential for electronic improvement to a particular process or activity, the analysis could provide the basis for identifying the relationships necessary for setting up a suitable system.

At this point it was obvious that there were some shortcomings in the manner of carrying out the analysis, but these were due to the research nature of the project and the consequent inexperience of the project team. There were no indications that the original hypothesis was incorrect (ie that the information requirements of an organisation could be derived by the progressive decomposition of a primary task model), and the study entered the next phase of deciding how the information base could be used in practice.

11.6 Using the Information

Having achieved a reasonably acceptable summary of information needs, there was a great deal of head scratching and heated discussion about how this could or should be used. The method is designed to help analysts to make enquiries about how well the information requirements are met in practice, so that areas of potential improvement can be highlighted. However, the size of the database resulted in some confusion as to the best way to proceed with the enquiry, and the next few stages were undertaken on a trial and error basis. The lessons learned from these trials are considered in the following paragraphs.

11.6.1 Activity Comparison

Even without rearranging the order of the records in the database, it was considered possible to use them to examine the actual organisation in a manner similar to that of the comparison stage of the SSM, ie by listing all activities and posing questions about their existence and effectiveness in the real situation, at the same time considering information-related problems. On this basis the earlier validation exercise was extended and further questions of the form *"When you carry out activity A, B, C etc, is this information available/ produced?"* were prepared.

The subsequent interviews reinforced the view that the detail of the information base was too fine in its original form to be the basis of searching questions, and some summation was found to be necessary to avoid redundancy, or posing questions that could insult the intelligence of the interviewee. For example, rather than cover item by item the information given to new residents and staff with questions such as *"Do you tell them where the post office/bed room/toilet/ etc is ?"*, it was necessary to ask the broader question of *"What information is given to newcomers?"*, resulting in a discussion about an informative brochure that everybody received on arrival at the EPH.

The process of asking questions about the whole range of EPH activities, although revealing a number of minor problems, was eventually discontinued because of the time involved in trying to cover all aspects of the database. After further experimentation, it became obvious that the analysis had to be more specifically targeted. This was achieved by utilising the sorting facilities of the database to focus the direction of the study on areas where there seemed to be potential for improvement.

11.6.2 Exploring by Type

It was subsequently agreed that the examination would address information needs by *type*, ie to concentrate on one or more categories where there appeared to be a problem, as indicated by the earlier interviews with staff and the issues reflected in the rich picture.

Sub-systems	System Status	Main Info Category	Problem	Comment	Priority
1.1	*Control*	COI	Poor definition of success factors	Being addressed by OIC	Medium
1.2	*Primary*	RI	Record-keeping/ care programmes	Worth investigating	High
1.3	*Support*	PI	Diverse info base	Outside local control	Low
1.4	*Support*	FI	Time-consuming record keeping	Worth investigating	High
1.5	*Secondary*	EI	Resident entitle-ment unclear	Outside local control	Low
1.6	*Secondary*	AI	Lack of info for staff deployment	Not significant	Low
1.7	*Secondary*	CI	No major problems	Could be improved, but low priority	Low
1.8	*Primary*	RI	Transfer of info between shifts	Manual improvements possible	Medium

Fig 11.14 - Extract From Priority Assessment Matrix

Accordingly, a report was prepared from the database listing all information by type, and discussions held with the OIC to determine a priority order for further study. As a guide to these priorities, a summary of the comments made about each type was prepared in matrix-format, showing also the main information category for each sub-system and the importance of each sub-system in relation to the primary role of the EPH (Fig 11.14). To prove the point that this is not an exact science, although the issues illustrated in the rich picture indicated some general concern about the need for more information on residents, leading to the conclusion that the RI category would be a good place to start, the OIC's main concern initially was the lack of accurate information about how to make sure that residents received their full entitlement of equipment (some W's can be more relevant than others - see Chapter 5!).

After further discussion it was agreed that, due to the complexities of the regulations surrounding financial transactions, and the obligation to make regular returns to the headquarters about financial matters, investigation of the FI category seemed to offer the most potential for improvement. Consequently, although the team did not wish to limit their investigations to one category only, it was agreed to examine financial matters first, and then review priorities for subsequent study.

Financial Information Category

To investigate the FI category, the database was first rearranged to separate out the FI items, which were then listed by functional group. This allowed a number of questions to be to be formulated from the activity lists (Fig 11.15) for discussion with all those persons in the EPH dealing with financial matters, the answers being noted on a separate formatted sheet.

Main Activity	Sub-Syst	Sub-activity	Related Information
1.4 Account for finance	1.4.1.1	Be aware of constraints	Budget allocation for EPH
			Budget headings for EPH
			County financial policy re EPHs
			Resident cash implications and constraints
	1.4.2.2	Establish & maintain accounts	How to establish & maintain accounts
			County accounting procedures

Fig 11.15 - Basis for Officer-related Questions

The questioning process was essentially undertaken in two stages; the analysis had indicated *what* information was required or produced, and it was first necessary to check whether this was the case in practice. If this was verified, further questions were posed about *how* the information appeared, and if it was acceptable. At either stage, if there was considered to be a problem, then it was noted for later discussion. For example, the analysis indicated that information about establishing and maintaining accounts was necessary in support of the finance function. If a related question (eg *"Have you received any training or instruction in setting up or keeping accounts"*) resulted in a negative answer, then this could be pursued further to find a way of improving the situation; for example by arranging for accounting instructions to be held at the EPH, or by ensuring that newcomers received formal or informal training in accounting procedures, and so on. However, if such information *was* available at the outset, further questions would be asked to determine if the form of the information could be improved by manual or other methods (Fig 11.16).

This exercise revealed that, although there were problems with the availability of information to support the finance function, generally these were outside the control of the OIC. For example, there was some confusion over a recent policy decision to allow OICs greater autonomy in the use and administration of their budgets, without any clear guidelines as to what this meant in practice, or when it would take effect. Most of the difficulties with the financial category were of this nature, and consequently not within the power of the defined problem-owner for this situation. Similarly, certain information *produced* by the finance function for external agencies (eg weekly returns of staff working hours for payment purposes) caused problems, but in most cases the form and frequency were already prescribed. Within the EPH, the necessary information about internal financial matters was readily available to authorised staff, and all associated records were held in a tidy and methodical manner. However, it was noted that the *processing* of financial data was laborious (eg budgetary calculations, recording the receipt and issue of residents' cash etc) and could possibly be improved in some way.

In other words, the exercise had revealed a number of difficulties associated with *what* information was needed or produced, but these were not under local control. With regard to *how* the information appeared in practice, there appeared to be little potential for improvement, except where information or data was subject to routine processing. It was noted that a number of forms and procedures had been devised by the Clerical Assistant and the OIC, and it was considered possible that the analysis might have been more revealing in other establishments where the approach to administrative work was not so methodical.

Resident Information Category

After further discussion, a similar exercise was carried out based on the RI category, with the results also indicating that most internal information storage and transfer needs were met within the EPH in an effective manner. As reflected in the original rich picture, there were concerns about the routine handover of information about the day-to-day condition of residents, and it was obvious that the *process* of keeping records for a large number of residents was an arduous manual task, indicating that some gains in efficiency might be possible by the introduction of computer support. Furthermore, when considering *what* information was produced for outside agencies, the aggregation of statistics on such matters as resident dependency levels seemed to offer some potential for improvement.

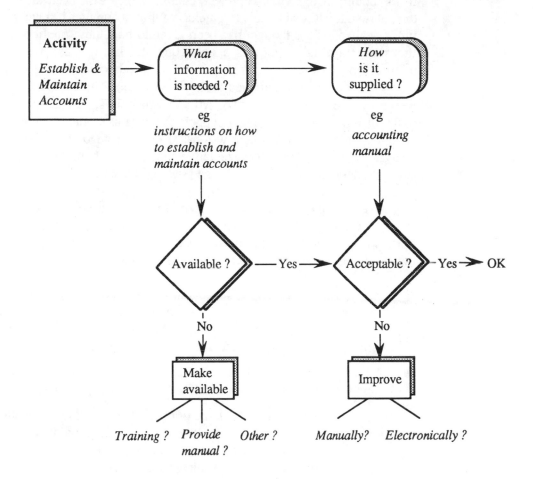

Fig 11.16 - Information Questioning

Other Information Categories

The FI and RI categories represented the largest proportion of information needs within the EPH, and specific question lists of the other categories were not prepared at this stage. However, the initial investigation had revealed that there was some potential for the use of a computer to support the primary functions (ie identifying and meeting resident needs), which in addition could be used to support other tasks, such as processing financial transactions and records. This was obviously an issue that could have affected other EPHs, and, after consultation with the OIC and the responsible officers at the department headquarters, it was agreed to pursue the development of a user requirement that would reflect the needs of this establishment and others within the local authority as a whole.

11.6.3 Development of a User Requirement

In the course of the MINSE project, comments had also been made to the effect that, although OICs were able to purchase a microcomputer if there were sufficient funds available, little thought had been given to how these might be applied, or to the staff training needs. As a consequence, limited use was made of the microcomputers that were already installed in other EPHs, and it appeared that many officers were dissatisfied with the situation. These views gave additional weight to the argument that the MINSE analysis should address the issue of defining how a computer could support the organisational objectives. The initial examination of the requirement was based on the categories derived in the earlier analysis, which produced some tentative conclusions about the type of records that could be held on a computer, and the potential advantages that could be gained. It was then necessary to review the work carried out so far, which, due to the extensive coding exercise and sheer size of the database, was making it difficult to 'see the wood for the trees'. Eventually it was realised that a number of other conclusions could be drawn from the analysis which were relevant to the development or selection of a computer system, such as:

- The activities required for successful operation of the EPH were known, and, based on the acquired knowledge of the EPH environment, those activities that were suitable for computer support could, in consultation with the client, be identified with reasonable accuracy.

- The information needed, or produced, by each activity was also known, and could form the basis of the records to be held on any computer system.

- If required, the activities could be prioritised on the basis of whether they were primary, secondary, support or control activities, providing the basis for the evaluation of any proposed systems. (This was a similar exercise to that carried out during the study of the Adult Training Centres described in Chapter 10, where a weighting was given to software depending on the relative value of the functions that it could support.)

In other words, by careful examination of the database it was possible to assess and list the processes that could be supported by a computer system, and identify broadly the associated information. As a result of this and an examination to collect volumetric data about the main EPH records and procedures, a draft user requirement was prepared, showing the potential for computer support and the likely costs of the options for system development.

Apart from the functional advantages, cost-justification was obviously an important consideration. The investigation had revealed that, due to low activity rates, an EPH would gain only limited benefits from computer support of isolated functions, and it was necessary to exploit the relationships that existed between the functions so that an integrated package could be developed. For instance the names of residents were used not only for personal records, but also for certain financial accounts, and for day-to-day tasks such as preparing menus and issuing medicines; the statistics about the dependency levels of residents were periodically collated from individual records and passed to the headquarters, and so on. The *source/ destination* codes added to the database during the MINSE analysis gave a fair indication of these relationships, and were used informally during the initial evaluation of software packages. This supported the earlier conclusion that the use of a more sophisticated software during the analysis phase would have been beneficial by allowing the relationships to be properly charted.

After the user requirement had been produced, and the options for system development had been considered, it was agreed that it would be most cost-effective if suitable applications software could be identified, which would also reduce the time required for installation. Initially, two application packages were examined, both of which had been developed for privately run EPHs and emphasised the medical aspects of residential care. Neither of these packages were entirely suitable for the operation of local authority establishments, and some customisation was considered necessary before the requirement could be met in full. However, to avoid unnecessary delays in meeting the expectations of the staff, and to keep development work and associated costs to a minimum, agreement was obtained to carry out pilot studies of these packages in two EPHs, ie the one involved in the MINSE project and another of similar size. As a result, a computer was installed in the EPH where the research was carried out shortly after the project was completed, which, being suitable for general administrative tasks such as wordprocessing, had the additional benefit of providing some immediate reward for the efforts of the staff who assisted with the analysis.

11.7 Conclusions

The primary aim of the project was to undertake research into the theory that the soft systems approach could be used to identify and explore the information needs of an organisation. In this respect the exercise was a valuable learning experience, and showed once again that there is sometimes a tremendous gap to be crossed between the theory and the practice of soft systems ideas. The analysts encountered many problems in the process of bridging this gap, but each problem was addressed in turn and an appropriate solution found, generally by making imaginative use of computer software to record and manipulate the mass of data that was produced as the analysis progressed. In retrospect, the research might have been easier in a situation with fewer functional elements, as a great deal of time and effort was directed at trying to tabulate the information links between the system components and construct a comprehensive information model. However, by applying the theory in the first instance to a complex situation the limitations of the software being used were highlighted, from which the obvious conclusions could be drawn. Even in these circumstances the approach provided a deep insight into the information needs of the organisation, and revealed a considerable number of potential improvements, some of which were implemented in the course of the study and others that were addressed in the longer term.

The most tangible outcome of the MINSE research project was the user requirement for computer support of EPHs, which, arguably, could have been produced using a more conventional approach. However, this resulted from an improved understanding of how certain information handling functions supported the organisation's objectives, rather than from a deliberate intention to introduce a computer to the situation. In the course of the exercise a great deal of knowledge had been gained about EPH activities and associated problems, and various other changes were made as a result of the investigation. For instance the responsibilities of the four officers were rearranged to reflect the structure of the first resolution conceptual model; the OIC personally addressed the issue of developing performance indicators for the system as a whole and the associated sub-systems; and measures were put in hand to improve the handover of information about residents between shifts. These latter actions highlight an important aspect of the MINSE approach, ie it is not necessarily concerned with changes of a technological nature, but with any that result in improvements to the problem situation; furthermore, because of the continual participation of the *people* that the organisation depends on, it helps to ensure a firm commitment to those changes.

In conclusion, the main findings from the project can be summarised as:

- A valid information model of an organisation can be compiled by adopting the MINSE approach.

- Full client participation is essential to ensure accuracy when compiling the information model.

- The model is generally used to assist with *enquiries* into the effectiveness of information provision, but could form the basis for an engineered system in appropriate circumstances (ie where a particular set of processes is being examined in detail).

- The approach encourages both manual and technological improvements to be considered.

- The manner of using the information model cannot be prescriptive, and will depend on the type of situation being studied

- Imaginative use of the sorting and report facilities of a flat-file database can facilitate the analysis process.

- It is not cost-effective to pursue the development of a full information model (ie showing communication links) using a flat-file database. However, the addition of simple codes indicating the general *source* or *destination* of information can be of value

- It is important that the analyst is aware of the *what/how* distinction in relation to information needs

- Like other soft systems studies, it is a *participative* approach and encourages full commitment to changes.

Additionally, the level of detail achieved by the extended breakdown of the conceptual model could have been utilised to explore other aspects of the formal systems model (eg resource levels, control or communication issues and so on), or to examine matters such as the ideal groupings of systemically desirable tasks, an approach discussed further in Chapter 12. As a final point it is worth reiterating that, although it takes a lot of time and intellectual effort to derive an acceptable information model using the MINSE approach, the model is independent of the actual form that the information takes in the real situation, and will remain valid until such time as the role of the organisation undergoes a fundamental change.

12 Investigating New Requirements

12.1 Introduction

It is not always possible to place soft systems studies into strict categories, such as those concerned with *organisational analysis, defining user requirements* and so on. Frequently the analyst does not set out with one of these specific aims in mind, but just applies systems thinking to generate ideas, making use of these ideas in whatever way proves to be of value. This same point was made earlier during the discussion of general applications, and examples were given of how it can assist with such things as planning, exploring relationships, putting particular aspects of an organisation into context, etc; in other words, using it as a basis for creative thinking and logical deduction. In a similar manner, it can often help to provide an overview of situations that are being affected by changes, particularly where these changes will cause a significant disturbance to the stability of the existing system. The effect of such disturbances on isolated parts of the organisation structure, individuals, resources etc may be obvious, but the overall effect and how the changes relate to current practices may not be so clear. Systems thinking can help to put all new and existing factors into context, clarifying relationships and providing a baseline for deciding what action is needed to absorb the new requirements.

This chapter examines a study, where, at the outset, the analysts were not entirely clear what could be achieved by using soft systems analysis, but felt it would provide an overview of the situation, enabling ideas to be formulated about new relationships that would arise from a fundamental change in role. The study was commissioned to examine the effect of the Education Reform Act (ERA) of 1988 on the non-teaching staff structures and responsibilities in certain Colleges of Further Education. In brief, the Act gives increased powers of delegation to Local Authority Schools and Colleges to manage their own affairs, with far greater autonomy than in previous years. It is the intention and the spirit of the Act to encourage them to be more accountable to their Boards of Governors, and, to a large extent, to become self-sufficient within the limits of their given finance and income generated by legitimate business activities. The Local Education Authority would continue to provide some central support and advisory services and be responsible for corporate matters concerning further education, but in most other respects, functional control would be passed to the individual establishments.

The Act was implemented in April 1990, and all the Colleges affected progressively took over their own affairs from that date. In the run-up to implementation there was naturally some concern, as it was obvious from the outset that changes would be required to accounting procedures and practices, personnel matters, and local control and management of resources. What was less obvious was the effect on the system as a whole, which had evolved over a long period of time influenced by a variety of political and economical pressures, and not necessarily in a methodical manner. A fresh look at the whole process was needed, and it was felt that this could be achieved by taking a *systems view,* unbiased by what was then happening in practice, ie before the Act took effect.

12.2 Overview of Approach

Although the brief for the study could be regarded as a *loose* one, the full SSM was not formally applied by moving deliberately through each of the SSM stages, although in retrospect there were many similarities. A form of rich picture was drawn up following the initial investigations, but it was used mainly as a summary of the structures, staffing levels and other essentially static factors that existed in the situation, and not as the basis for selecting a relevant system. The initial conceptual models were developed almost intuitively, based on the idea that, for a College to be self-sufficient, the income received must be at least equal to the cost of providing the further education courses at the required level. This concept is illustrated by the *equality paradigm* shown in Fig 12.1, where the resources required to provide further education have been accurately defined and costed, the necessary finance obtained and used to purchase the resources, which in turn satisfy the needs for education.

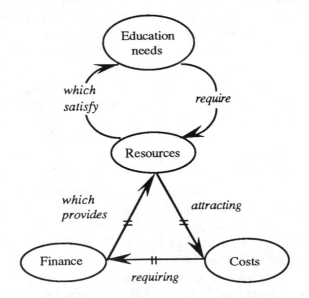

Fig 12.1 - The Equality Paradigm

From this starting point, a series of models were developed to determine what changes in practices would encourage Colleges to move towards a *state of equilibrium* in the new circumstances. To do this, an ideal model based on the equality paradigm was constructed and then compared with a neutral model reflecting the further education system as it existed before the Act took effect. These two models were compared, leading to a third model that showed what could happen once the Act was implemented. It was hoped that this would reveal the differences to be taken into account when establishing a revised organisation structure, and also highlight the areas of opportunity to enable the Colleges to move towards the state implied by the ideal model. The approach, which is explained in more detail in the following sections, is summarised in Fig 12.2.

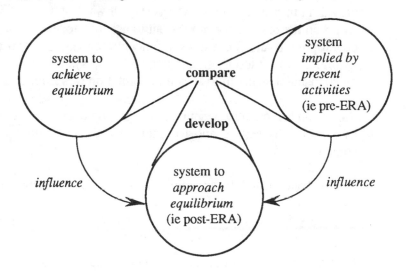

Fig 12.2 - Summary of Approach

12.2.1 Responsibilities and Roles

To put the discussion into context, responsibility for providing further education in the United Kingdom is divided between three hierarchical levels, with the central government Department of Education and Science having overall responsibility, which is devolved to Local Education Authorities (LEAs) each accountable for the education provision in specified geographical areas. *Delivery* of the appropriate courses is the responsibility of specified Colleges, in accordance with the plan for the area; prior to implementation of the ERA, each was controlled in respect of staffing levels and general mode of operation by the parent LEA.

The majority of College finance is provided by central government grants, which are divided at local level into that required to provide central administrative and support services, and that used to finance the College operations, allocated on a pro-rata basis to individual establishments dependent on the level of course provision and student numbers. This forms the bulk of the College *budget,* which can be supplemented by income generated by additional courses provided to meet specific demands, which in effect are outside the statutory requirements.

12.3 The System Models

The models illustrated in this chapter are based on those used during presentations to the staff of the Colleges involved in the studies. As such, they are simplified versions of the working models used during the analysis, which became extremely complicated due to the interactivity between the components, usually represented as a series of inputs and outputs of relevant information. They are also the end-product of a prolonged exercise where a variety of root definitions and conceptual models were developed, then discarded or amended as the analysis progressed. The starting point was a description of a system that reflected the equality paradigm, ie showing the activities necessary to maintain equilibrium when providing further education. At this stage it was not the intention to explore the real world on the basis of the model, but to use the model as a starting point for further development, and the analysts concentrated on specifying the transformation required, without too much concern about the system ownership, actors, etc. After much discussion, the following transformation was agreed:

'to obtain and organise resources at appropriate levels to deliver Further Education as required by statute and to meet local demand, unconstrained by financial considerations'

On this basis, a model was progressively developed that would:

a. Allow finance to be equated to the resources needed to provide education at the required level (ie achieve equilibrium)

b. Organise these resources to provide the appropriate courses.

Within this ideal model (Fig 12.3), the actual costs of providing education at the appropriate level would be calculated for both academic and non-academic resources (eg staff, equipment, facilities, services, etc). On this basis the necessary funding would be obtained for ongoing conversion to resources at the correct levels. Once resources have been obtained, they would be organised to provide the College curriculum, and ensure the necessary non-teaching support.

Although expressed simply, it is assumed that the necessary liaison takes place between the sub-systems of the model on matters of common interest. In systems terms, there would also need to be mechanisms for taking control or corrective action based on defined measures of performance, related in this case to the effectiveness of education provision (ie ensuring that the curriculum delivered meets the defined education requirements), and to the accurate provision of resources.

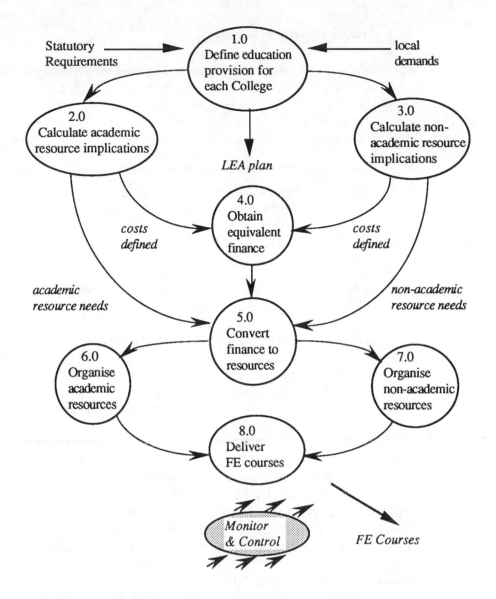

Fig 12.3 - Model of Ideal System

It was realised, of course, that this ideal could never be achieved or sustained in practice, as precise information on resource needs is not easy to obtain, therefore accurate costs cannot be calculated. Furthermore, even if it *were* possible to make the necessary calculations, the finance available is unlikely to be equivalent to these costs. Consequently, resource levels may never fully satisfy the needs, and an imbalance between academic and non-academic staff, equipment, services, etc is inevitable. However, the ideal could be pursued, which was the basis of the arguments that were considered at a later stage.

12.3.1 Pre-ERA System Model

The next step was to relate this model to what was then happening in practice, ie before the Act took effect. This was achieved by developing a neutral model, based on the following CATWOE components:

C   -   recipients or students of further education

A   -   staff of the LEA headquarters, and College teaching/non-teaching staff

T   -   obtain and organise resources to deliver education at the appropriate level to meet statutory requirements and local demand

W   -   a *primary task* system

O   -   the LEA (ie on an *ownership hierarchy* basis)

E   -   influenced by central (ie headquarters) requirement to ensure parity of all Colleges in the geographical area of responsibility

The second model (Fig 12.4) illustrates a systemic view of the pre-ERA situation, assuming that, although the primary aim was to deliver education to meet statutory requirements and local needs, the provision of the necessary resources was constrained by limitations on finance, in turn influenced by the requirement to ensure parity of resources in all Colleges controlled by the LEA.

Consequently, the model excludes activities associated with making precise calculations of *actual* resource needs and costs, and reflects the use of weighted factors to determine instead the *relative* costs of courses, thus allowing the available finance to be allocated on an equitable basis to Colleges, and within Colleges, to Departments. This provides the College budget, which, based on the local detailed plan for course provision, is converted into resources, constrained not only by finance, but also by central directives about staffing levels and grades, the use of support services etc, and the way the support services are organised and controlled. The role of the Colleges in this situation was limited to one of *service management*, ie organisers and providers of the prescribed education using resources allocated by the LEA headquarters, with little flexibility to enhance levels or types of resource to meet shortfalls. The division between the headquarters and the College responsibilities is shown as occurring between sub-systems *4.0* and *5.0* in the system model.

In this situation, where the fundamental requirement is to allocate available finance on an equitable basis to Colleges, it is unlikely that the system can achieve equilibrium. However, prior to the Act taking effect, Colleges were effectively underwritten by the LEA, and an input of additional finance could normally be expected in the event of major shortfalls, compensating for any underlying inequality in the system. The removal of this insurance was, of course, a significant factor in terms of developments, as the consequences of poor management in the future would have to be addressed at local level.

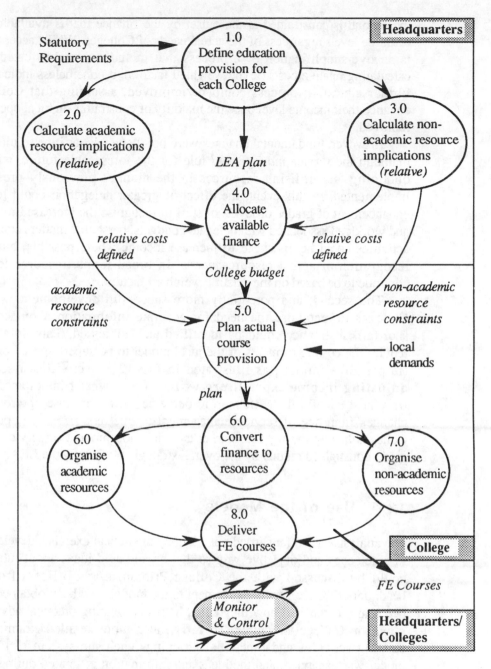

Fig 12.4 - Pre-ERA Model

12.3.2 Post-ERA Models

There are certain obvious differences between the first two models, the most significant being the basis of their construction, ie the first is *demand driven,* relies on an accurate assessment of costs, and is not concerned with financial constraints;

the second is constrained at the outset by the finance that *is* available, restricting resource levels regardless of the actual needs of Colleges. Whilst accepting that this is an oversimplification of a process that includes a sophisticated method of calculating relative costs (ie the weighted factors), it nonetheless indicated how the further education process could be improved, assuming that Colleges could enhance their income level once the majority of constraints on local operations were removed.

However, fundamental changes were not inevitable, as it was quite possible to maintain the *service management* role for the foreseeable future, with Colleges obtaining and utilising resources in the manner previously prescribed and implemented. In this event, the effect of greater delegation could lead to some enhancement of grades or levels of staff to recognise the increased responsibilities and workload associated with delegated budgets, personnel matters etc; commercial activities could be enhanced to increase income where possible, but not in true recognition of actual resource needs. In broad terms, resource levels would continue to be based on the existing weighted factors.

The second, and potentially more successful development, would be for Colleges to recognise the need for accurate information about costs, so that shortfalls in finance could be identified and addressed. This, in effect, would require the costing elements of the ideal model to be superimposed on the current (ie pre-ERA model), as illustrated in Fig 12.5. Activities associated with **adjusting income/expenditure** would take on greater importance, and would be related to defined shortfalls in resource needs. In this case, the role effectively changes from one of service management to one of a *franchise* operation, with Colleges having the right to market the education product or service in a defined areas, constrained mainly by the overall strategic plan set by the LEA.

12.4 Use of the Models

The analysts had so far carried out a largely theoretical exercise, developing models in a detached and objective way so that broad-based ideas about future strategies could be discussed with the College Principals and officers from the LEA headquarters. Once general agreement to the models had been obtained, it was then necessary to consider how they could help in providing practical advice about the effect on College structures. To carry out a more detailed examination of the existing activities and functional groupings each sub-system of the post-ERA model was expanded, and the ideas and information generated during the exercise used to explore the real situation. Although not explicitly regarded as such at the time, this equated closely to the *comparison* stage of the SSM, as described in the following sections.

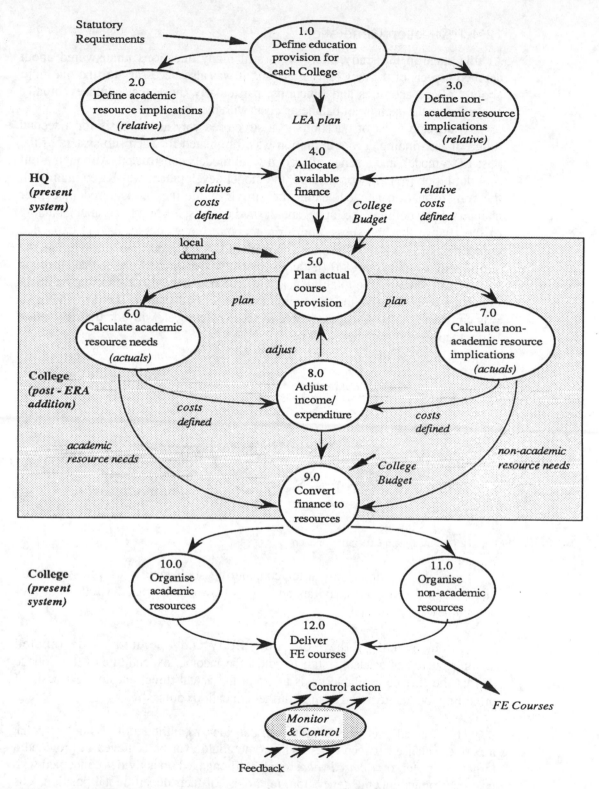

Statutory Requirements

1.0 Define education provision for each College

LEA plan

2.0 Define academic resource implications *(relative)*

3.0 Define non-academic resource implications *(relative)*

HQ *(present system)*

4.0 Allocate available finance

relative costs defined

relative costs defined

College Budget

local demand

5.0 Plan actual course provision

plan

plan

6.0 Calculate academic resource needs *(actuals)*

7.0 Calculate non-academic resource implications *(actuals)*

College *(post - ERA addition)*

adjust

8.0 Adjust income/ expenditure

costs defined

costs defined

academic resource needs

non-academic resource needs

College Budget

9.0 Convert finance to resources

College *(present system)*

10.0 Organise academic resources

11.0 Organise non-academic resources

12.0 Deliver FE courses

Control action

Monitor & Control

Feedback

FE Courses

Fig 12.5 - Post ERA Model

12.4.1 Question Generation

At this stage in the study, there were still many questions unanswered about detailed aspects of the new legislation, and it was also necessary to find out more about College activities and functions, particularly those that could eventually contribute to the exercise of balancing costs with income.

To develop a list of questions in a structured way as the basis for a second round of fact-finding, a root definition was formulated for each sub-system of the post-ERA model, and a series of second level models constructed. Although a full root definition for each was derived, the model development was based mainly on the transformation needed, using the simple guidelines considered in earlier chapters, ie by defining the inputs and desired outputs, leading to an understanding of the change that the sub-system brings about. For example, having expanded components *1.0* to *7.0* of the model shown in Fig 12.5, it became apparent that the main input to sub-system *8.0* (ie 'adjust income/expenditure') was the detail of the College allocation of the total budget, with secondary inputs of information about the actual costs of providing the appropriate level of education. The desired output was considered to be *finance adjusted* to the correct level, with the transformation being the link between the two (Fig 12.6).

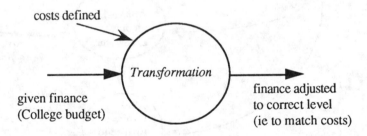

costs defined

Transformation

given finance
(College budget)

finance adjusted
to correct level
(ie to match costs)

Fig 12.6 - Sub-system 8.0 Transformation

From this starting point, after considering such factors as who would be involved in the implied activities and system ownership etc, the following root definition was formulated:

'A system owned by the Governing Body and operated by designated administrative and academic staff to adjust the income/ expenditure of the College so that the finance available tends to equal the actual direct and indirect costs of providing the statutory and desired further education courses'

This seemed a reasonable system to consider, as, although it was unlikely that a perfect balance between costs and income could ever be achieved, it provided a target to aim for, or a *shared value* for all staff engaged on activities concerned with cost control and income generation, regardless of their departmental position. The main environmental factors were taken to be the effect of market forces on such things as staff recruitment, purchase of materials, and course demand, etc, constrained by the long-term contractual obligations to existing teaching and non-teaching staff. The model constructed from this definition is shown in Fig 12.7.

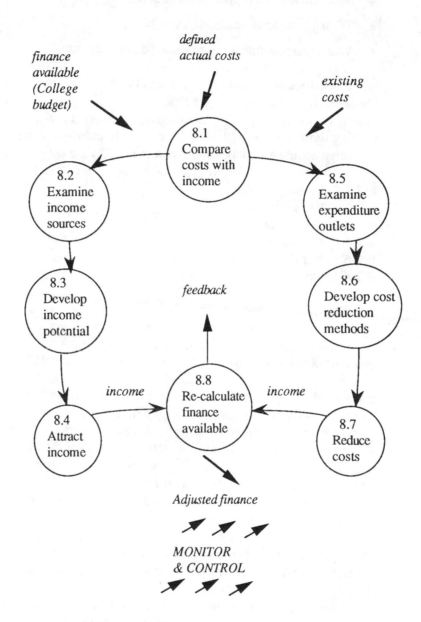

Fig 12.7 - Expansion of Sub-system 8.0

Examination of this and the other expanded models enabled a series of questions to be prepared to aid the analysis, eg:

- What cost-control methods currently exist ?

- What are the main expenditure outlets ?

- What are the total income sources ?

- Who is responsible for purchase of equipment, etc ?

It also focussed attention at each level on the measures of performance required for effective monitoring and control, and the information dependencies of all sub-systems. Further expansion was carried out on the same basis, progressively moving through the model until more specific activities started to emerge, such as those that could be associated with component *8.7*, ie *reduce costs* :

8.7.1   Monitor and control
8.7.2   Reduce waste
8.7.3   Tender for services
8.7.4   Control purchase
8.7.5   Introduce improved methods
8.7.6   Consider new technology

This further expansion gave greater insight into the information requirements of the system (eg for the activity *Tender for services*, information about tender procedures, potential contractors, standards required, etc could be relevant), and could be checked against the information actually available in practice, an approach that is examined further in Chapter 11. Analysis to this level was not undertaken for all sub-systems, only those where some significant change was implied, as even the secondary expansion was extensive and time-consuming. This was mainly because of the multi-functional nature of a College, which, in addition to the many activities associated directly with course provision and support, also provides a full range of student services. As an example of the level of detail produced at the secondary modelling level, the outputs from the sub-systems concerned with calculating academic and non-academic resource needs included lists of:

Direct Support Resources - Lecturers, classroom technicians, classrooms, workshops and laboratories, equipment, facilities, and so on.

Indirect Support Resources - Administrative and clerical staff, central support technicians, cleaning and catering staff, grounds and building maintenance implications, etc.

The questions raised during the modelling exercise were a mixture of those concerned with furthering the knowledge of the analysts in relation to the facts about present procedures and responsibilities, and those concerned with whether or not the systemically desirable activities were being undertaken, by whom, and with what level of effectiveness. This mixture of basic fact-finding and the *comparison stage of the SSM* eventually allowed some conclusions to be drawn about the possible effect of the new legislation on existing groups and responsibilities.

12.4.2 Effect on Functional Groupings

Although the analysis was of necessity extensive, it was not the intention of the study to explore each activity in detail to determine where procedural or other changes were required, but to provide the basis for reviewing existing structures, particularly those related to non-teaching staff. As with all systems exercises, because the model does not reflect functional groupings that have evolved for administrative convenience or other reasons, making the link between the model and the organisation was problematic. To do this in the College study, each activity drawn out from the model was itemised, and a judgment made as to who, or which functional group, might be responsible for carrying out the task if it existed. This posed the complementary questions of 'do the activities exist' and 'who could be carrying them out', leading to the normal enquiries about effectiveness and efficiency.

It is worth noting that imaginative use of unsophisticated technology (as discussed in Chapter 7) can aid this part of an investigation, enabling a large volume of activities that may be undertaken by specified functional groups to be readily summarised. In the case of the Colleges, the increased delegation of responsibilities bought about by new legislation warranted the introduction of new endeavours, or changes to existing ones, and the analysis helped to identify where these were required at all levels. Each activity was assessed in relation to the *function* that it could form part of, and whether a *functional group* currently existed that could be responsible for it. Where no such group existed, it implied that some addition to the organisation was needed; in other cases, the new requirements indicated a change in the responsibilities of existing functional groups. This process is summarised in Fig 12.8, and is essentially a variation on the Checkland matrix, customised to suit the particular circumstances.

Bearing in mind that the review was directed at determining the broad changes necessary to a College structure, at this point the analysts were primarily concerned with those areas where new functions were required. For example, the **management information function** referred to frequently in the chart did not exist in most Colleges before the study was carried out, but appeared to be essential for effective operation in the future. The lower order changes in responsibility were considered at a later stage when working with Principals to determine the detailed changes required at individual Colleges.

Sub-System	Example of Activities	Assessed Function	Assessed Responsibility	Exist or Not	Global Grouping
5.0 Plan actual course provision	Interpret Local Education Plan Decide other courses Make decisions about provision Monitor & Control	Policy-making Strategic planning Management Establish Management Information mechanisms	Governing body Management Team Mgt Information Function	Yes Yes No	Executive
6.0 Calculate academic resource needs	Assess student demand Assess student/lecturer ratios Assess equipment, classroom, facilities etc Monitor & Control	Planning Personnel Resource allocation Management Information	Heads of teaching departments Staffing Administration Mgt Information Function	Yes Yes Yes No	Operational Support Support Support ?
7.0 Calculate non-academic resource needs	Determine type & levels of staff Assess equipment/facilities Assess maintenance/service needs Calculate costs Monitor & Control	Personnel Resource allocation Maintenance/building services Finance Management Information	Staffing Administration Finance Mgt Information Function	Yes Yes Yes No	Support Support Support Support ?
8.0 Adjust income & expenditure	Examine income sources Develop income potential Examine expenditure outlets Reduce costs Monitor & Control	Marketing/sales Cost-control Management Information	Business Function Finance Mgt Information Function	No Yes No	*Not known* Support Support?
9.0 Convert finance to resources	Establish & operate accounts Tender for services Recruit staff Purchase equipment etc Pay bills etc Monitor & Control	Finance Contracts Personnel Purchasing Finance Management Information	Finance Administration Staffing Administration Finance Mgt Information Function	Yes Yes Yes Yes Yes No	Support Support Support Support Support Support
10. Organise academic resources	Assess lecturer availability Allocate direct non-teaching staff Prepare timetables Produce curriculum Monitor & Control	Academic resource allocation Course planning Management Information	Heads of teaching departments Mgt Information Function	Yes Yes No	Operations Operations Support ?

Fig 12.8 - Tabular Summary of Activity/Function Relationship

Cross-relating each activity from the model to the existing functions indicated that a number of these could remain virtually unchanged, in particular those concerned with organising academic resources and delivering courses. In generic terms, these were regarded as the **operational** functions of the College, ie those primarily concerned with delivering the end product. Similarly, many of the activities could be grouped under a generic **support services** heading, ie those concerned with organising indirect services such as finance, personnel and general administration etc.

In broad terms, the College organisation structures prior to the Act reflected this two-pronged approach (ie operational and support). Within this structure, the legislation would require an enhancement of the finance and personnel functions, and recognition of the increased responsibilities of College Administrative Officers, especially in relation to the servicing of the new Governing Bodies. The analysis findings are summarised in Fig 12.9, showing new functions that could be needed in the changed circumstances, but weren't formally included in existing structures, eg:

a. Business and marketing functions, ie those required to develop income potential.

b. Cost and purchasing control.

c. Coordination or control of management information systems, inter alia, to provide the basis for calculating the cost of resources, and establishing effective monitoring and control mechanisms throughout.

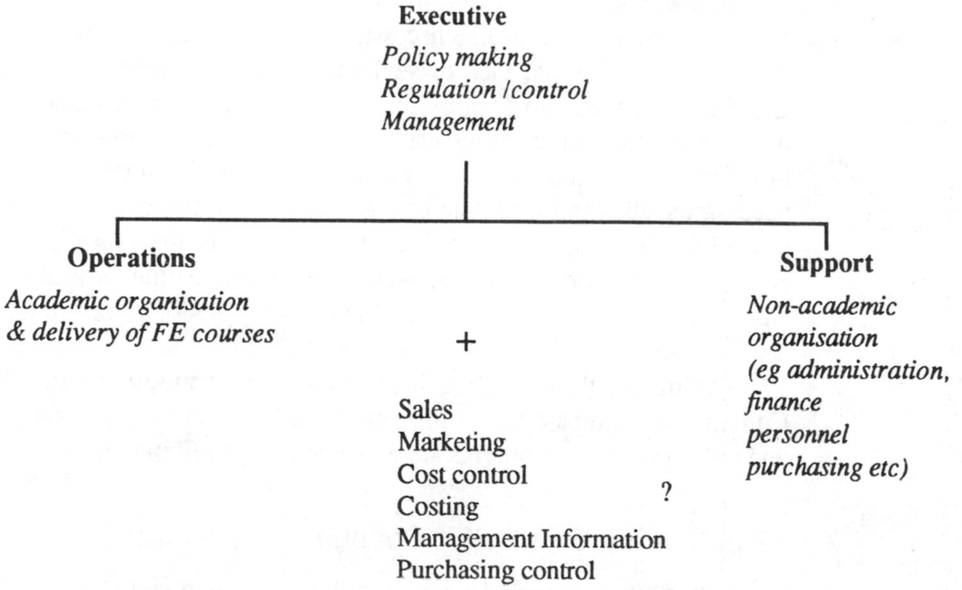

Executive

Policy making
Regulation / control
Management

Operations

Academic organisation
& delivery of FE courses

+

Sales
Marketing
Cost control
Costing
Management Information
Purchasing control

?

Support

Non-academic
organisation
(eg administration,
finance
personnel
purchasing etc)

Fig 12.9 - Summary of Analysis Findings

This approach also gave rise to a change in attitudes about the split between the academic and non-academic functions within the Colleges, which encouraged the Principals to make more flexible use of resources. Traditionally, all non-teaching staff had been grouped together regardless of their actual duties, but the concept of **operational** and **support** groups (rather than *teaching and non-teaching*) suggested that some of these staff who were directly involved with classroom activities (eg workshop technicians) could reasonably come under the control of the lecturing department heads, recognising their direct contribution to the delivery of courses and involvement in the *operational* side of the business.

12.5 Conclusions

Following acceptance by the client of the findings of the systems exercise, and their value for focussing the direction of College developments, it was then necessary to provide some pragmatic advice about the new grades, salaries and groupings that would be required once the Act was implemented. It was accepted from the outset of the study that it would be unrealistic to expect changes to be made overnight, and that a gradual evolution to the desired state would be necessary. The final part of the exercise, therefore, was to merge the systems conclusions with a more conventional examination of the various real-life factors affecting the way in which new groupings could be achieved. For example, there was a continuing requirement to comply with the pay scales used by the Authority as a whole, although the Governing Bodies were free to appoint new staff at any level in the scale. There were natural concerns about ensuring a degree of parity of gradings and structures between and within Colleges, not least to avoid industrial relations disputes; it was also necessary to take due account of the frameworks and levels of expertise that then existed and evolve a strategy for change, rather than take radical action. Most important, of course, was the recognition that, although in the long-term measures to increase income and reduce costs would enable Colleges to become more self-sufficient, in the short-term the finance available would not necessarily allow all desirable changes to be introduced simultaneously, and some prioritising would be required. Furthermore, the analysts were aware of major computer developments that were taking place that would eventually have significant effects on the duties and responsibilities of certain functional groups at the Colleges.

Taking all these factors into account, a strategy for the development of Colleges was prepared, and discussed with the client organisation. This strategy and the supportive arguments are explained briefly in the following paragraphs

12.5.1 Requirements for Change

The system models represented an ideal that applied equally to the situation before and after the implementation of the ERA. However, given the increased powers of delegation to Colleges authorised by the Act, and the effective removal of certain constraints in terms of staffing and budgetary control, Colleges will have more freedom in the future to make adjustments locally to resource levels. The analysis showed that there were many similarities between the activities that are systemically desirable and those undertaken at the time of the study, but highlighted certain aspects of the organisation that needed strengthening to allow the Colleges to become more self-supporting once the ERA took effect. Overall, changes could be required to *functions* and to the *structure* as a whole.

Functional Developments

- More emphasis would need to be placed on calculating the actual costs of providing FE courses, so there was a baseline for deciding what adjustments to expenditure and income would be required. Although there were already many activities concerned with cost and financial control, these functions would assume greater importance in the future, inferring that the expertise in Colleges should be developed or enhanced, and officers nominated to assume new responsibilities. To this end, the analysts also examined the procedures and responsibilities for financial matters at one specific College, taking account of the earlier systems findings and the forthcoming introduction of computer support.

- Similarly, Colleges were involved with income-generating and marketing activities to some extent. After the Act took effect, it would become increasingly important to coordinate the efforts of all personnel in this respect, and to establish a focal point for business activities. This could be achieved in a number of ways, for example by establishing a business centre or by nominating one person as head of business services, at the discretion of individual establishments.

- Management information systems would need to be developed. Given the varying levels of computer expertise in Colleges at that time, and the current status of the proposed new systems, it was difficult to make specific proposals about how the associated responsibilities should be handled at College level. The term *management information systems* itself covers a wide variety of, mainly, computer developments, ranging from the installation of stand-alone microcomputers and wordprocessors, to the sophistication of the proposed computer system. The importance of providing a focal point for these developments was also emphasised, and it was felt that this could be part of the role of the College Administration Officer so that non-partisan strategies for computer developments could be evolved.

- To recognise the increased responsibilities that Colleges would assume for finance, personnel and local administrative matters, the related functions that existed before the Act took effect would need strengthening.

Structural Changes

The system model, although not intended to reflect ideal functional groupings, also indicated that a sub-division of activities to those directly concerned with course delivery (ie the operational functions), and those concerned with non-teaching tasks (ie the support functions), was appropriate to align systemically-related activities. This tended to support the broad structure at most Colleges at that time, ie reflecting a division to operations and support. There were, however, certain functions grouped under the generic support heading that were involved in the delivery of courses (eg departmental assistants and technicians), and Colleges were advised to consider their position in the organisation structure in light of the expressed argument. Similarly, the library and student services functions, which also related to the delivery of courses, could be regarded as part of the operations group.

In high-level terms, cost-control and management information functions seemed to fit comfortably within the generic support heading. However, activities concerned with the development of income sources (eg business functions), whilst dependent on the administrative services for day-to-day support and information, and overlapping the business activities of some teaching departments, are primarily concerned with the interface between the College and potential income suppliers, eg public and private sector employers, government agencies etc, and the public. Arguably, these functions could be located within either the support or operational groupings, or be established as a separate agency with liaison links as appropriate.

In addition, the analysis indicated that each of the main functional divisions (ie operational, support and business) should be represented at executive level so that all aspects of College activities would be reflected in the development of policies and in the overall management of the College. Representation of the support functions would be accomplished by nominating a 'Support Executive' from those officers who would be responsible for the revised finance, personnel or administrative groups.

12.5.2 Strategy for Change

Whilst recognising the need to introduce measures that would make Colleges more self-supporting, it was considered unrealistic to expect such measures to be effective immediately. In terms of priorities, the client was advised to consider first the changes required to reflect the responsibilities that would be assumed from April 1st 1990, accepting that, in the longer term, further changes would be needed as a result of management information and computer system developments, and to improve cost control and income-generating activities. Although all Colleges that were within the control of the LEA had broadly similar organisation structures, at a detailed level there were significant differences (ie in structural arrangements and in the numbers and expertise of staff), and it was not appropriate to make detailed proposals for individual Colleges at this stage.

Taking all these factors into account, the analysts advised that the following actions by College Principals were appropriate in the short-term, in consultation with the LEA as required:

a. Examine the existing administrative and clerical responsibilities to determine how the new functional groupings of finance, personnel, and administration could be achieved, together with support service representation at executive level. As part of this examination, the requirement to provide a focal point for management information development should also be considered.

b. Determine the grades to be applied for revised posts that would assume responsibility for all financial and cost-control matters, for personnel, and for enhanced administrative support.

c. Determine the local strategy for filling these posts, taking account of the existing expertise within the College.

d. Review the effect of these changes in relation to administrative and clerical staff overall, at the same time considering the potential advantages of reallocating those posts that provide direct support for teaching departments.

It was also agreed that, once the changes implied by these actions were effective in each College, further measures would be taken to reflect the computer developments and to address the issue of adjusting finance in relation to actual costs (ie by enhancing cost-control and income-generating activities) so that the Colleges could move towards the state of *financial equilibrium* implied by the system models.

12.6 Summary

Following the systems exercise and a full report to the client, the structures at two Colleges were examined in detail and revisions made to reflect the new requirements. The ideas expressed in the report, and the example structures, were then used to assist the remaining Colleges within the control of the LEA in the development of their own strategies for change.

The approach used during the study did not follow strictly any prescribed format, but nonetheless made effective use of the soft systems ideas to produce results that were well received by the Principals of the participating Colleges. The process of decomposing system models mirrored the requirement to consider first the overall effect of the Act on functional groupings and structures, then to examine responsibilities and tasks in some detail, ie taking a *top-down* approach that was initially unconstrained by existing factors but progressively took them into account. Significantly, it allowed the relationships between existing activities and those considered necessary in the future to be more clearly understood, illustrating the benefit of using the soft systems approach in a proactive role.

13 Procedure Audit

13.1 Introduction

Procedure Audit is a particular application of soft systems ideas developed by the Northern Region of the Institute of Management Services (IMS) Technical Board, with the main authors being Peter Mcloughlin of The New College, Durham, and Geoff Brown of Sedgefield District Council. The description that follows is based on an article that appeared in the November 1989 issue of the IMS Journal, and additional material supplied by the authors. It provides an invaluable example of the way in which soft systems ideas can be of benefit to MS professionals and other analysts, and, in turn, to those organisations that make use of their services.

Most practitioners undertake reviews of procedures at some time or other, whether they are procedures in the commonly accepted sense of the term, ie low-level processes or activities with a *set order of steps* (Collins Standard Reference Dictionary), or the total activities of an organisation that combine to achieve the purpose of the business as a whole. Procedures of both sorts can be found in any type or size of organisation, for filling in forms, processing goods, inputting data to a computer system etc, and those directed at the continual development and implementation of policy, or the preservation of the mode of business operation. Procedure Audit can be used to examine and review any of these, not as a financial tool or a means of apportioning blame for failure, but as a device for assessing the effectiveness of a procedure and the efficient use of associated resources.

It is not a work or method study exercise, but a way of establishing and monitoring agreed measures of performance, as a basis for determining whether action needs to be taken to bring this performance within desired parameters. The practitioner or analyst is not necessarily concerned with implementing changes, but with identifying where such changes could be beneficial so that appropriate action can be taken. In this respect and others the approach has similarities with the SSM, particularly with regards to client participation to encourage a commitment to any changes that result from the audit.

13.2 The Rationale of Procedure Audit

A procedure can be viewed in systems terms, ie transforming inputs to outputs to achieve a desired end (Fig 13.1). The inputs could be details entered onto a form, with a completed form as the output, or the raw materials that are fed into a production process for conversion to assembled components.

Fig 13.1 - Procedure In Systems Terms

A procedure carried out by *people* with the aid of tools technology etc can therefore be considered as a Human Activity System, one that is comparable with the formal systems model described in Chapter 3. Ideally, it will reflect the characteristics of the model, eg comprise a series of interrelated sub-systems, have resources for its own use, an expectation of continuity, and be part of a wider system that is influenced by the surrounding environment. It should also have defined measures of performance, a mechanism for recognising when deviations occur, and a means of taking control action when necessary, factors that are particularly relevant to the Procedure Audit process.

Like other Human Activity Systems, procedures can be considered from a number of viewpoints, each leading to the derivation of different criteria for success. For example, typing a document may be a time-consuming chore for a secretary who might view success in terms of how quickly the chore can be completed, but the author will be concerned with the accuracy of the conversion from manuscript to type, and the value of the final document as a means of communication. These concepts are utilised in the Procedure Audit process to determine measures of success from a specified point of view, as agreed by the person responsible for the procedure and its continued existence, which in practical terms is generally the person who can authorise changes to be made.

This approach also encourages the analyst to look beyond the basic procedure to the wider systems that it is part of and the environment that can affect both requirements and performance, in the process gaining an understanding of all the relevant factors that need to be taken into account. Taking a series of views across the whole situation, referred to as the *hill tops approach* by Mcloughlin and Brown, recognises that the wider system could be instrumental in:

- setting the objectives or defining the purpose of the procedure

- influencing the person making decisions as part of the procedure

- monitoring or measuring performance of the procedure

- providing the resources to carry it out

In practice, this means that the analyst needs to take account of such factors as the rules, regulations, and criteria for performance laid down by the parent organisation. These may consist of constraints imposed by the finance section, the legal or tax advisors, those who set the house-style for documents, and so on, representing elements of the wider system, which is influenced in turn by environmental factors. A procedure does not exist in isolation from its surroundings, and Procedure Audit requires this to be acknowledged explicitly. These ideas are summarised by Mcloughlin and Brown in a manner similar to that shown in Fig 13.2.

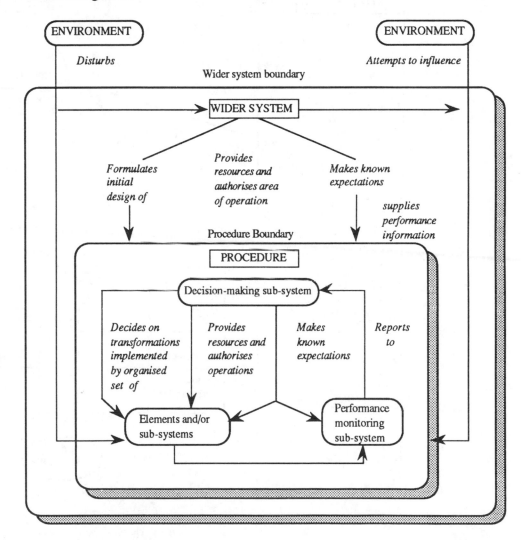

Fig 13.2 - Procedures and Wider Systems

This is a derivation of the formal systems model, and is used in essentially the same way, ie as a checklist to ensure that the model of a procedure is well-formulated and all the necessary attributes are considered during the analysis.

13.2.1 Procedure Audit and Conceptual Modelling

Procedure Audit follows the same process as the SSM, using a conceptual model of the defined human activity to make a comparison with what is actually happening in practice. The approach also takes into account that the way objectives and goals are defined can be influenced by the perceptions, attitudes and behaviour of individuals.

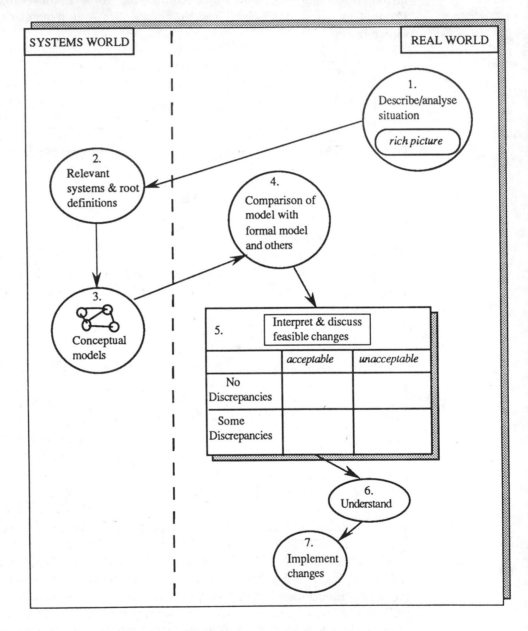

Fig 13.3 - Summary of Procedure Audit Conceptual Approach

In any situation where there is felt to be problems, these will not exist independently of the human beings involved. The perception of the problem will differ, depending on such things as the views and experiences of individuals, and their beliefs about what is or what ought to be. Furthermore, the analyst must realise that problems do not exist in isolation, but are interrelated with other problems affecting the same situation. It is important, therefore, that all the circumstances are considered during a Procedure Audit, including peoples' views of problems and potential solutions. Involvement of individuals in the development of these solutions will also make it easier to implement any changes that are necessary. A conceptual model is constructed to take these factors into account, and used to make a comparison with the existing situation.

The rationale underpinning Procedure Audit, therefore, is that a conceptual model of a chosen procedure can be constructed on similar lines to that of the SSM, which, after checking with the formal systems model, can be compared with the real situation to determine where changes are appropriate, taking account of defined measures of performance. This variation of the SSM is summarised in Fig 13.3.

13.3   The Practice of Procedure Audit

In practical terms, the audit process can be summarised as a sequence of stages (Fig 13.4) although, like the SSM, these should be regarded as guidelines rather than prescriptive, as there may be a need to refer back to earlier activities and make adjustments as the audit progresses.

Stage 1 - Create Commitment

The first step toward putting these ideas into practice is to gain a commitment from the client organisation to establish and support the audit mechanism, and where changes are warranted, to make these changes effective. The extent to which this is required, and the difficulty in achieving it, will depend on the level of procedure that the audit is directed at. A major review of the business operation (a *meta-procedure?*), which could comprise a series of nested procedures, will require commitment at all levels. In this instance participation and team-building are particularly important, together with the development of objectives to which all individuals, sections, departments and so on can contribute. For lower-level activities involving fewer staff, commitment is probably easier to achieve, but is no less important in ensuring that the audit process is effective.

It is also crucial to avoid any suggestion of recrimination, regardless of the level of procedure being examined. The audit is essentially aimed at identifying and rectifying procedural flaws, and not at the individual's performance, although realistically this may feature when considering efficiency aspects at a later stage. Once again, involvement of staff in the process can be of value in overcoming these fears, encouraging a feeling of association with any problems identified and their solutions.

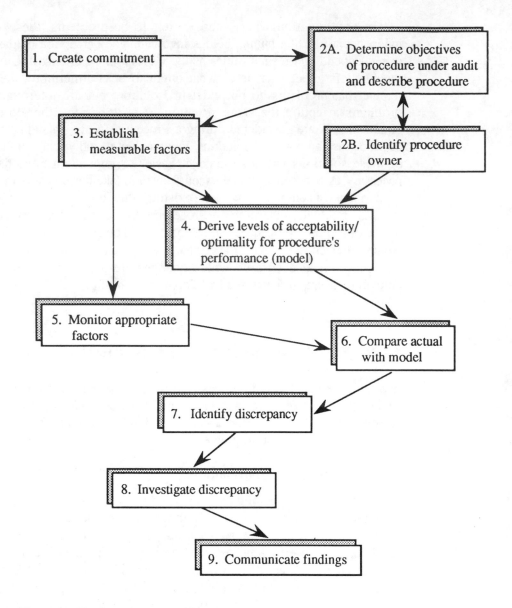

Fig 13.4 - The Procedure Audit Stages

Stages 2a & 2b - Determine the Objectives and Owner of the Procedure

The ownership and objectives of a procedure interrelate, as each will vary with the viewpoint taken, and both these factors need to be considered before the measures of performance can be derived. The analyst must aim to define an uncontentious view of the transformation the procedure aims to achieve, bearing in mind the variety of viewpoints that could be considered relevant. The example in Fig 13.5 shows a replenishment procedure for a stationery store, viewed by a user in terms of the timescale of delivery, quality of the order, and potential deficiencies, whilst the storeman has subjective feelings about his own status, wages, and so on.

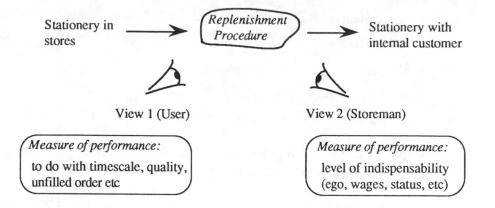

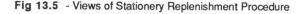

Fig 13.5 - Views of Stationery Replenishment Procedure

Different viewpoints lead to different models which, in turn require different measures of performance, and it is necessary to ensure that the model used as the basis for the audit is one that reflects the *organisation's* needs in relation to the procedure. Having agreed this with the procedure owner, a *procedure description* is prepared, showing amongst other things, the resources and activities needed to meet the objectives, and this is compared with what actually exists. Significant discrepancies between the two may be resolved at this stage, ie before monitoring is carried out. The activities are then decomposed until such time as the related resources (eg people, time, money, skills etc) which need to be committed to the procedure can be clarified. Any deviation from this in practice would be taken as a constraint, the analysis having revealed where such constraints are necessary so they can be taken into account when carrying out the audit.

Stage 3 - Establish Measurable Factors

As with other types of soft systems exercises, deciding performance measures is of prime importance, but can prove extremely difficult, particularly when considering abstract ideas such as customer satisfaction. Asking questions about the consequences or signs of failure can be of assistance at this stage. For example, posing the question *"What happens if the stationery replenishment procedure fails?"* could give rise to a variety of answers. The customer might suggest:

- *"I get the wrong amount of stationery"*
- *"I get stationery at the wrong time"*
- *"I cannot get the correct stationery"*

and so on, from which a statement of a performance measure could be derived logically, ie the procedure is effective if *the customer receives the right amount of the correct stationery at the right time.*

Alternatively, if the storeman's view were to be considered, the answers might be:

- *"I get complaints from the customer"*
- *"My boss thinks I'm inefficient"*
- *"I won't get a pay rise next year"*

indicating that the procedure is effective if *the storeman remains in good standing with both the customers and the boss, and receives pay rises when eligible.*

The latter is, of course, unlikely to be chosen and would be extremely difficult to measure, but does emphasise the need to determine a neutral view of a procedure before attempting to measure its success.

Stage 4 - Determine Levels of Acceptability or Optimality

Once the performance measures have been agreed, the acceptable or optimal levels of achievement have to be determined. It may be possible to gain access to past records so that a realistic assessment can be made of achievable levels; alternatively, in the absence of any historical data, utopian levels (eg 100%) could be set initially, then modified if the auditing process highlights that they are unattainable. The earlier decomposition could reveal activities within the procedure that are amenable to monitoring (such as adherence to bin-card re-ordering procedures), which could obviously have an effect on the success of the procedure as a whole. When auditing *meta-procedures*, if the overall exercise revealed failings, these lower-order activities would also need monitoring to determine more precisely the reasons for this failure.

Stage 5 - Monitor Appropriate Indicators

At this stage mechanisms for collecting information about attainment levels need to be designed and implemented. Mcloughlin and Brown emphasise that this exercise must be undertaken after due consultation with all those involved, recognising that such matters as customer satisfaction cannot be addressed without the involvement of the customer, who could be asked to score the quality of the service provided. This may require additional information to be recorded on requisition forms etc, or involve setting up new registers, all of which will demand extra staff time, and could lead to some resistance. The underlying purpose of recording new information must be made clear to avoid it being regarded as just a bureaucratic exercise.

Stage 6 - Comparison of Levels

For comparison purposes, there are fundamentally two choices, ie either the attainment levels can be assessed against an agreed target, or relative comparisons can be made, eg against what happened in the past if records are available, or what is happening at the start of the exercise compared with what happens at selected future points in time, effectively establishing **better** or **worse** levels, as summarised in Fig 13.6.

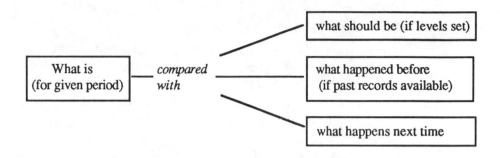

Fig 13.6 - Comparison of Levels

Stage 7 - Identify and Evaluate Discrepancy

The comparison stage should produce one of the following results:

1. Under- achievement, ie worse than target level

2. Over- achievement, ie better than target level

3. No discrepancy between levels

Where there is a difference in levels, either as an under or over achievement, then some investigation is required. Even for over achievement, or where there appears to be no problem, it could be worth reviewing the target levels to consider if these have been set incorrectly; perfection the first time round might well be suspect! Once the analyst and the client are satisfied that the procedure is effective, then it is necessary to decide if it is also efficient, as effectiveness could be achieved by employing unnecessary resources. It is also worth noting that efficiency should not be considered until the effectiveness is proven, the procedure could be doing the wrong thing in an extremely efficient manner.

Stage 8 - Investigate the Discrepancy

At the start of the exercise, the actual procedure was checked against a conceptual model that included the activities deemed necessary to achieve the agreed transformation. If the changes revealed by this comparison have not been made prior to the audit commencing, this could be the reason for the failure to meet the target levels of attainment. If, however, the procedure does comply with the procedure description derived earlier, then it will be necessary to subject each element or activity of the model to further scrutiny, for example by breaking them down to lower-order activities, or by auditing each element in the same manner as the parent procedure. This process is similar to work study exercises where measurable elements of a task are identified before the actual measurement takes place, and consideration is given to the resources required at each level, including the skills and experience of the operator.

Stage 9 - Communicate Findings

The final stage is to communicate the findings of the audit to the procedure owner, to allow decisions to be made about the action to be taken. To improve the presentation of results a **Procedure Audit Report** may be devised, listing the purpose and objectives of the procedure, the nested activities and required resources, the performance indicators and desirable target levels, and the results of the analysis.

At this point, the auditing process is essentially completed, the role of the analyst in this context being an investigatory one, rather than as an agent of the change. In practice the brief for the study may well include a mandate for advising on the implementation of the changes, and, in certain circumstances, actually installing revised procedures. The distinction, nonetheless, needs to be made, so that the aim of Procedure Audit is properly understood, ie to identify levels of disorder within procedures.

13.4 Procedure Audit - An Example Application

To provide a summary of these stages, and give an example of how Procedure Audit might work in practice, consider typical procedures for claiming travel and other expenses in a small section of a large organisation. This process frequently comprises the following steps:

- Individuals informally record expenses incurred (eg in notebooks etc).

- Pro-format claim forms are completed at the end of a set period, eg one month.

- The claim forms for each section are checked and totalled by administrative assistants, and totals transferred to section registers for record purposes.

After leaving the section, the completed forms are passed to the department personnel section for verification of the claimants eligibility, and then processed by the data preparation section of the Finance department onto the main computer for subsequent issue of cheques. For this exercise, the procedure being examined is that carried out within a section, and the procedure owner is taken to be the person who could authorise changes at this level (eg by the introduction of pro-formats for day-to-day recording of mileage etc), in this case the head of each section.

13.4.1 Establishing the Audit Criteria

Firstly, to explore the idea that procedures do not exist in isolation, the relationships between the internal procedure, the wider system and the environment can be considered (Fig 13.7).

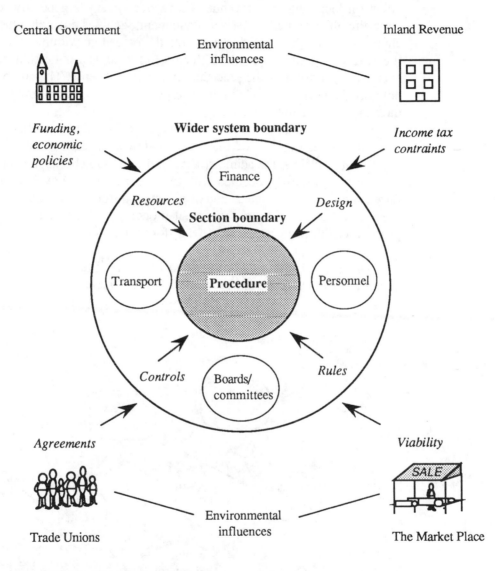

Fig 13.7 - The Wider System & Environment

The wider system would include the parent department, the personnel section who verify the claims, the finance department that influences the format and timing of claims, etc, and the boards or committees that determine the organisation policy as a whole.

These elements of the wider system are in turn influenced by factors of the surrounding environment, for example the liability for income tax imposed by the Inland Revenue, trade union agreements, market trading conditions, and, for public service organisations in particular, the policies of central government, and initiatives such as the introduction of the community charge and the subsequent effect on local authority revenue. The wider system is generally responsible for allocating the resources of people, equipment, facilities, including the design of the claim forms which are used to complete the procedure within each section; setting the rules governing the eligibility of individuals for certain expenses; and exercising control over the procedure in the widest sense. The environment affects not only the decisions made in these respects, but also the amount of expenses that the company can afford to pay.

Secondly, a number of views of this procedure can be considered relevant, eg from the point of view of the individual who is concerned with the speedy settlement of claims, the administrative assistant who checks the claims for errors, and the boss who is concerned with the effect on his status if a large number of false claims were to be made, and discovered outside the section. An uncontentious view could give rise to a conceptual model (Fig 13.8) that reflects the actual procedures followed, based on the transformation:

'to collect and transfer information to the finance department about legitimate expenses incurred in the course of normal employment'

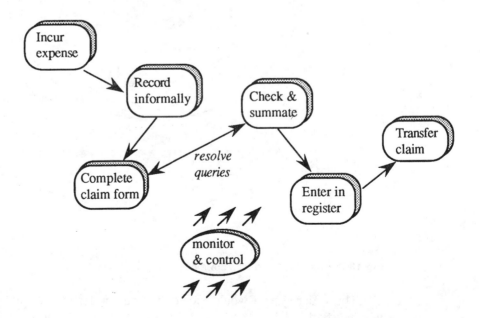

Fig 13.8 - Neutral Conceptual Model of Expense Claim Procedure

By examining each of the nested activities, a list of appropriate resources can be prepared, including:

- *time of individual claimants*

- *notebooks etc*

- *claim forms*

- *time of administrative assistant*

- *calculator*

- *section register*

- *information about rules and regulations*

For this simplistic example, it can be assumed that these are readily available, and no adjustment of the resources is needed before the audit process can start. The next step is to consider whether the notional system reflects the characteristics of the formal system model. The existence of the wider systems and the environment have been explicitly considered, and at this stage are mainly relevant to ensuring that the analysts is aware of the whole situation and the influencing factors. At a later stage, if errors are found in the procedure's performance which cannot be traced to a failing within the local system, it may be necessary to examine these influencing factors further, for instance the design of the form might be inadequate for the accurate recording of information, a factor outside the direct control of the section involved. The resources required by the procedure have been identified, and some degree of interconnection between the envisaged sub-systems is apparent, mainly involving the transfer of information. (It is worth noting that Mcloughlin and Brown also make use of other general system models eg *control, communication* and *human factor models,* during their investigations.)

The procedure owner at this level is the section head, who will have the authority to act as the regulator of the procedure's performance. It is then necessary to develop appropriate measures for this performance. To illustrate the effect of taking different viewpoints, if the views of the main actors in the situation were to be considered, performance measures such as the following could be derived:

Individual claimant - *the speed of receiving the expenses cheque.*

Administrative assistant - *the accuracy of the checking procedure, ie the number of errors undetected at this stage that could be referred back by the personnel section or the finance department.*

Section head - *the number of occasions that fraudulent claims are made, ie indicators of honesty or dishonesty.*

In relation to the primary task model, one acceptable indicator of success in achieving the transformation (ie collecting and transferring information about legitimate expenses) could be the occurrence of queries, either from the claimant with regard to the expenses received, reflecting the accuracy of the information transferred, or from the recipients of the claim forms, eg the personnel section and the finance department. In terms of target levels, because of the importance to the individual of receiving the correct level of payment, then 100% attainment would probably be aimed for.

13.4.2 Monitoring the Procedure

To test this particular application of Procedure Audit, a monitoring process was established in a section of thirty persons, and the administrative assistant briefed to record the number of queries referred back over a period of two months. During the first month, only one query was raised, in this case by the personnel section, but it was considered important enough to warrant some corrective action.

A claim had been made for travel outside the normal operating area, a journey which should have been authorised before it took place. On investigation, it was apparent that the claimant had been unaware of the regulations in this respect, and they had also been overlooked by the supervisor and administrative assistant who were involved in processing the claim. It was accepted that there had been no deliberate attempt to claim for unauthorised travel, but it was obvious that the regulations had not been publicised effectively, and the section head took steps to ensure that they were more widely circulated to prevent any further occurrences. This incident was an isolated one, but it served to highlight the overlap between the possible measures of performance that had been considered; an undetected error not only slowed down the payment of the expenses, but could also have bought into question the honesty of the individual and thus affected the status of the section head.

During the second month, two unsigned claims were processed and passed for payment, which were then returned to the claimants for signature. As a consequence, the claims missed the normal weekly batch-processing of data into the organisation's mainframe computer, and the subsequent delay in receiving the expenses cheques was greater than the actual time taken to reprocess the forms within the section. Although this was a relatively minor error, it indicated that the process for checking the claims (ie a *sub-system* of the main procedure) should be improved so that expenses would be received in the most timely manner.

The most significant outcome of this relatively simple test was an increased awareness within the section of the effects of not adhering closely to the ideal requirements of the procedure (as reflected in the conceptual model). As a result, a greater commitment to accurate recording was made by all the staff concerned, generally improving the *effectiveness* of the procedure overall. No further changes were considered necessary to address *efficiency* aspects, as, in this instance, the activity rate was low and there were few resources involved. It is worth bearing in mind that, in terms of the wider system, the data on the forms was subject to a number of validation checks, and failure of the procedure at section level would not have had any disastrous consequences. However, the exercise highlighted the value of the Procedure Audit approach for not only identifying procedural flaws, but also for encouraging the commitment of staff to making improvements effective.

13.5 Conclusions

Procedure Audit has been thoroughly tested across a wide range of situations by Mcloughlin and Brown, and can be applied at any level within an organisation. For illustrative purposes, only a simple example has been given here; however, the approach has been used for a number of major investigations, including a study of the procedures of a local authority for making policy decisions about housing stock, covering in excess of fifteen thousand buildings. (Hitherto, information about these and other studies has not been published, but they are covered by the regular workshops held by the IMS to promote the approach.) At whatever level Procedure Audit is applied, it is a worthy addition to the toolboxes of all analysts or managers who are concerned at any time with making improvements of a procedural nature, and provides a further example of how soft systems ideas can be used to good effect for everyday analysis work.

14 Bridging the Pragmademic Gap

14.1 Introduction

This chapter picks up on the development of the FAOR methodology, examining the concluding phase where all the main components, including the SSM, were tested in the field as a final trial of the package as a whole. The package is used to explore a given situation, identify a problem area, then produce a requirements specification for a computer system to effect some improvement. The description, illustrations, and conclusions given in pages 231 to 249 are based on the FAOR publication, which provides an excellent example of the use of the SSM to generate productive ideas about a situation, and to do so in an ordered and logical manner. The work that followed to finalise the user requirement is hitherto unpublished, and is derived from a series of discussion papers raised during the latter stages of the project.

In this chapter I have also endeavoured to clarify the relationship between the soft systems approach and other more conventional methods of examination. In the opening pages of the book I referred to the *pragmademic gap* that seems to exist between pragmatists who have not had the benefit of a formal systems education, but might wish to make use of soft systems ideas, and those who have developed and applied these ideas at a more academic level. The term could also be used to describe the gap between the *practice* of solving problems, and the somewhat *academic* exercise of systems thinking, where ideas can be generated about desirable changes that, in light of subsequent investigation, cannot be realised fully in the real-world.

To develop this theme, two distinct phases of a study into the information support needs of an organisation are examined here, the first making extensive use of the SSM to obtain an overview of these needs and produce a high-level user requirement for computer support, and the second applying standard O&M techniques to progress the study to an acceptable conclusion. These exercises were essentially discrete studies in their own right, and, whilst certain conclusions can be drawn about the complementary nature of the different approaches, taken overall, the study also highlights the importance of merging *systems thinking* with *real-world* considerations.

14.1.1 Background to the Study

Until this study was carried out the individual elements of the FAOR methodology had been developed and tested in isolation, each being the responsibility of different organisations within the FAOR partnership. The purpose of the final case study was to test the viability of the package as a whole, and all the main partners were represented in the team that addressed the selected problem situation. Unlike the exercise described in Chapter 10, the team did not include representatives from the parent organisation (ie the County Council), a deliberate omission to avoid the results being conditioned by the knowledge or views of internal analysts, who nonetheless assisted with the general arrangements and administration. Since the earlier study, the team's experience with the SSM had increased considerably, and any difficulties encountered at this stage were related more to the use of the FAOR package as an integrated whole than to the individual components themselves.

14.1.2 The Client Organisation

The chosen organisation was a local authority that had participated in earlier developments, but this time the study examined the problems that had arisen in the *Highways* and the *Planning* departments following the establishment of a common unit providing both with financial, administrative, personnel and computer support. It was felt that this unit had not met the expectations of the department heads, and some improvements were necessary, without a clear definition of where these might be achieved. This was considered to be a scenario ideally suited for the soft systems element of the FAOR approach, and it was hoped that the exercise would not only reveal where the problems lay, but also produce a requirements specification to show how technology could provide solutions.

The Highways and Planning departments are separate organisations within the local authority, each under the control of a specialist Chief Officer and with a range of responsibilities, including:

Highways:

- construction of new roads, and the design of major road schemes and bridges
- maintenance of existing roads
- traffic management and road safety
- street lighting
- refuse disposal sites

Planning:

- environmental services, including countryside design and development
- control of local plans and development
- co-ordination of public transport
- promotion of employment opportunities

Some of these responsibilities were undertaken as agents of the central government Department of Transport, with agency arrangements also in force with District Councils in the corresponding administrative area.

The departments were co-located in the same building, and, as it was felt that their roles were similar and some economy of scale would be achieved in terms of staffing, equipment, and accommodation, a Central Services Support Unit (CSSU) had been established. This included a Registry function for receipt and despatch of correspondence, acting also as custodian of the central filing system, which was supplemented in the branches of the two departments by outposted files maintained by branch clerks. The CSSU, which became the main focus of the study, together with a centralised forward planning unit, employed a total of 150 staff under the joint control of the two Chief Officers.

14.2 Summary of FAOR Activities

The FAOR methodology was described briefly in Chapter 9, and for ease of reference the main components of the package are shown again in Fig 14.1.

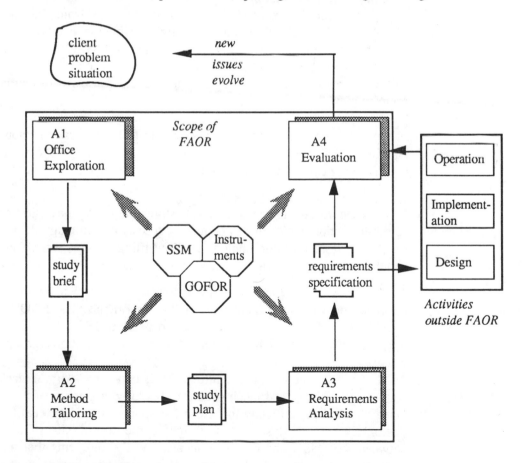

Fig 14.1 - The FAOR Methodology

It is not the purpose of this book to give an in-depth explanation of the methodology, as the theory and practice of the approach is covered extensively in the FAOR publication. However, in order to demonstrate how it makes use of the SSM to complement the other components of the package, its application during the case study is summarised briefly in the following sections.

14.2.1 Activity 1 - Office Exploration

The FAOR methodology is directed at analysing the requirements of offices for information technology improvements, and the first *Activity* is to carry out an exploration of the chosen office domain, using the SSM as required. Apart from familiarising the analyst with the particular problem situation, it is also concerned with identifying those aspects of the organisation where there may be issues that need addressing. During the field trial, this exploration was carried out in a conventional manner, eg by gathering background information relevant to the client organisation, by becoming familiar with the physical layout of the offices involved, and by interviewing a representative sample of staff from both departments and from the central support unit.

Rich pictures were used to express the findings from the interviews, then combined to form global pictures showing the situation as a whole. One of these, showing the relationships between the central *service providers* and the *information users* throughout the Departments is shown in Fig 14.2. Following the analysis, a number of issues emerged that seemed worthy of further consideration, for example:

Information usage: this was problematic due to the diversity of sources, such as planning applications, conservation areas, highways design specifications, and so on, involving communication with external bodies as well as other departments of the Council.

Information storage: there was a need to store a large volume of accurate and up-to-date information, which had to be readily accessible. Although there was a facility for central storage of information, localised filing systems were also in operation, and there appeared to be a large amount of duplication.

Communication problems: there also appeared to be problems with communication between and within the branches of both departments. The FAOR team felt this could cause difficulties where information of mutual interest was being handled, information which could have benefited people in all functional groups.

In addition, the team noted further issues concerned with the lack of an office automation strategy and incompatible technology, a paucity of resource tools for managers, and the difficulty that central support staff had in identifying with their customers. The analysis also highlighted the complexity of the relationship between the CSSU and the departments, and the problems that inevitably arise when trying to meet the demands of a large number of customers whose requirements differ. Amongst other things, the examination had revealed possible areas where improvements could be made to the services that the CSSU provided, in particular those services concerned with supplying the information needed for staff to fulfil their functions.

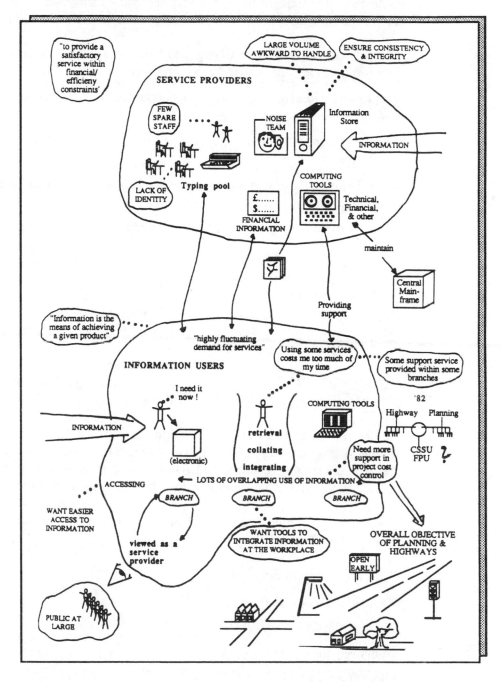

Fig 14.2 - Rich Picture of Highways and Planning Situation (courtesy of Schafer et al, <u>Functional Analysis of Office Requirements</u>, Wiley 1988)

Defining Relevant Systems

Having expressed the main factors about the situation in rich picture form, the analysts than considered viewpoints that were relevant, and the systems that might be developed to explore the issues that had been uncovered. At this point in the study the following systems were selected as relevant to the situation :

- An **information handling system** to support the manipulation of information and the needs of individual users.

- A **communication system** to support lateral communication, both within and outside the departments.

- An **information management system** to manage the distribution of information.

- An **information storage system** for efficient storage of large volumes of information.

- An **information retrieval system** for rapid access to information.

Tentative root definitions for each of these were prepared, and it was then necessary to discuss them with the client to obtain feedback on the analysts' ideas, and to consider the direction that the study would take. (The involvement of the client in the selection of relevant systems is stressed in Chapter 8.) The discussions indicated that the client was mainly concerned with improving the speed of access to information and the efficiency of information storage. Consequently, the draft root definitions for the storage and retrieval systems were combined to form a new definition for an **Information Provision System** (generally referred to as the IPS). This included the following transformations:

".....to collect appropriate information, as defined by the activities of the Highways and Planning departments, to maintain that information cohesively, such that it is consistent, accurate, up-to-date, complete and easy to manage, and to provide access to that information within the time constraints of the users, ..."

A conceptual model constructed from this definition is illustrated in Fig 14.3, showing the main activities of:

Collecting information

Maintaining information

Making information available to users

Monitoring the system

Allocating resources

Deciding criteria for assessing performance of the systems

Following the sequence of the FAOR methodology, the output from Activity 1 is a statement of the study brief, and this was prepared as a result of these discussions and the modelling exercises. It was agreed, therefore, that the main focus of the study would be *to investigate the information provision and access requirements of the Highways and Planning departments ...., with a view to improving the level and quality of the service, within current resource constraints.*

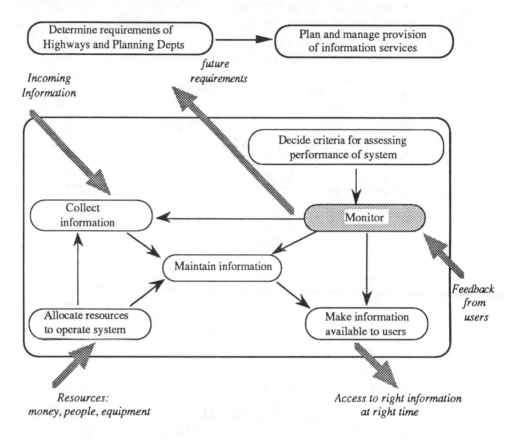

Fig 14.3 - Conceptual Model of Information Provision System

14.2.2 Activity 2 - Method Tailoring

The second *Activity* is concerned with tailoring the method to suit the particular circumstances, and the FAOR package introduces the concept of **instruments**, each representing a pre-packaged set of tools, techniques and application principles which can be used as required to address a situation from a desired viewpoint. In the Method Tailoring phase, the analyst considers the type of instrument, or instruments, that are considered suitable for the particular investigation and decides their specific role in the analysis. The instruments provided and their agreed use in the case study are summarised in the paragraphs that follow.

Functional Analysis Instrument (FAI)

The FAI is designed to assist with the task of understanding a given office as a functional system within the organisation. In brief, it is used to identify, define and characterise the essential functions that the office serves in order to fulfil the organisational objectives. This allows a planned office system to be perceived in terms of these functions, and to specify clearly what should be achieved by implementing any new system. In relation to the development of the IPS, the FAI was used to define the areas of the organisation that were concerned with information provision. In practice this meant clarifying the roles of the service providers and service users to determine which processes would require support from the proposed system. The FAI was also used to define the scope of the other instruments.

Communications Analysis Instrument (CAI)

The CAI is employed to examine the office as a communication system, where information objects (such as documents, files, microform, etc) are transferred between individuals and functional groups in support of the organisational objectives. In particular, this instrument was used during the case study to examine the information needs of customers in terms of time, security, reliability etc, and to decide how these factors might be improved.

Information Analysis Instrument (IAI)

The function of this instrument is to address information objects and their form and usage. It was utilised to explore the characteristics of the information objects to be covered by the IPS, eg determining the form of the information to suit the users' needs, and assessing whether or not the objects that then existed were adequate.

Needs Analysis Instrument (NAI)

As explained in Chapter 8, the NAI addresses the needs of *people;* in the context of the FAOR package, it is concerned with clarifying the interests and preferences of those who will use any designed office system, and how they could affect its success. During the study, it was also used to address some of the issues uncovered during the first Activity, particularly with regard to cooperation and coordination.

Benefits Analysis Instrument (BAI)

The BAI, which is used within a *Benefits Analysis Framework,* provides concepts, criteria and fundamental steps to aid the evaluation of potential benefits to the users of new office systems. After the IPS requirements had been specified during Activity 3 of the FAOR approach, it was employed to forecast the possible effects of the proposed system in terms of benefits and disadvantages (the latter being referred to *disbenefits* in the FAOR publication).

14.2.3 Activity 3 - Analysis

During *Activity 3,* more detailed analysis is undertaken by applying selected instruments to the problem situation. It is generally accepted that an analysis of functions (ie using the FAI) is a prerequisite to the use of the other instruments, as illustrated in 14.4, which shows the planned sequence of application for the study; moving from the method tailoring phase to the development of the requirements specification, followed by an evaluation of the proposed system in the final stages using the Benefits Analysis Instrument.

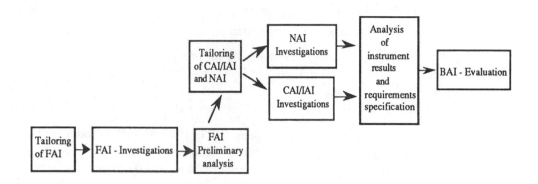

Fig 14.4 - Sequence of Investigations in Activity 3

The analysts were concerned at this point with improving their understanding, by taking the conceptual model of the IPS as a starting point, and using the instruments to explore the relationships between the model and the actual situation. This was carried out over a three-week period using a mixture of interviews and questionnaires covering a representative sample of persons from the CSSU and the departments. To assess the needs of the users in relation to the support actually provided, the team interviewed the service providers and the users within each branch of the departments.

Functional Analysis Conclusions

The wide diversity of tasks carried out by the Departments made it inappropriate to carry out a detailed examination of each function. However, the functional analysis enabled areas of commonality to be identified at a high or general level, showing how and where the two departments interacted. For example, both were concerned to some extent with the following:

- The production of the infrastructure plan for the area covered by the local authority

- The design of detailed changes to the plan

- The control of design changes

- The control and execution of changes to the plan

- The maintenance of the infrastructure

The Planning department was mainly concerned with County-wide planning and design aspects, and the Highways department with the detailed design of specific structures such as roads and bridges, including control and implementation of changes. *Design* and *control* appeared to be the main areas of commonality, indicating that a high degree of cooperation and communication was needed with regard to information of mutual interest. Some general conclusions were also drawn in terms of the requirements for the IPS, ie that :

1. It must support the technical staff (ie engineers) and their clerks alike, and be integrated with the departments.

2. The major emphasis must be on providing the users with an overview of the available information, and how the information is interrelated.

Information Analysis & Communication Analysis - Conclusions

The analysis of the information objects and communication aspects, carried out as a joint exercise to ensure method economy, was undertaken in two stages, ie first by interviewing a few staff, then distributing a questionnaire based on the initial findings. Questions were posed about matters such as the perceived importance of certain information objects, the most frequent communication means and partners, and the perceived problems (and possible solutions) relating to information objects and the current services. The subsequent analysis enabled certain conclusions to be drawn about the importance of knowing about, and having access to, current information; the availability of material; the low utilisation of the CSSU Registry; and the need for alternative methods for organising and locating information. Fig 14.5 shows the type of information objects used by the Departments and the main issues regarding information handling and communication.

Needs Analysis - Conclusions

The introduction of computer support could have had significantly different effects on the *service providers* and on the *users,* and the Needs Analysis was carried out using a different questionnaire for each generic group. Service providers were questioned about such matters as their attitudes towards new technology, and their willingness to adapt to new circumstances, whilst the emphasis for service users was on the possible effect in terms of work results and achievements. Amongst other things, the Needs Analysis indicated that:

- Users were unsure of the level of service to expect, and there was uncertainty among the service providers about how the service was utilised and the criteria that should determine quality.

- The tasks of the CSSU staff were poorly defined, and it was not possible to achieve a high level of identification with the organisation.

- There was a readiness to accept new technology and a general desire for improvements.

- The majority of interviewees were not well informed about the services provided by the CSSU, and, partly as a consequence, the services were considered to be inadequate.

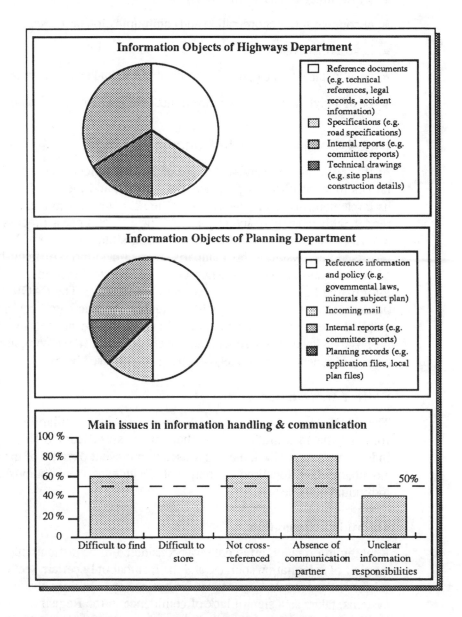

Fig 14.5 - Example Results of Information and Communication Analysis

The role of the Needs Analysis is essentially to reveal user-related issues that should be taken into account when producing the requirements specification in the latter stages of a study. In this case the analysis indicated that, in terms of users' needs, the proposed IPS should:

- allow the work of the CSSU to be more clearly definable

- improve contact between the service providers and users

- allow direct access by customers

- encourage better coordination and communication processes

- have a high availability

- be adjustable to the particular needs of individual technical branches

- be of a high quality, eg in terms of accuracy and cross-references

14.2.4 Comparison with Conceptual Model

At this point in the exercise, the use of the instruments had revealed a number of factors in terms of the *requirements* of the IPS, and the *problems* that could affect its implementation and effectiveness. The next step was to compare the conceptual model with what was actually happening in practice, a process similar to that described in Chapter 9, where the existing situation is considered in light of a proposed new system. This comparison was made more difficult by the existence of two organisations with a number of disparate functions, each using similar information but stored and utilised in different ways. The FAOR team noted these differences, but unfortunately did not realise the true significance in terms of the detailed user requirement that would eventually be needed to progress the IPS proposals. However, these variances were reflected in the comparison findings for each component of the model, which are summarised below.

Collect Information

The FAOR team felt that, although there were some similarities in the sources and form of the information used, there were significant differences in the way indexes were organised, the codes used to represent districts of the County, and the extent to which localised (ie non-central) indexes were established and used by individual branches.

Maintain Information

Keeping information up to date was considered to be difficult due to the increasing number of information sources, and the fact that only certain sections of documents might be superseded by new information. In addition, the use of local filing systems, taken as a sign of lack of confidence in the Registry, led to duplication of items of common interest, and, pragmatically, inefficient utilisation of office accommodation for storing files. Localised filing being more difficult to control than central systems, there was evidence of out-dated files being held in office space, which exacerbated storage problems. It was noted by the team that, although this was tolerable in the accommodation then occupied, a forthcoming move of both departments to a new central complex could make the volume of files difficult to accommodate, office space being more restricted. The problems were not confined to standard paper-based files; the work of the departments required a variety of file forms, such as large plans, maps and drawings, and computer storage media.

Make Information Available

Examination of the activities in the real situation indicated a number of possible problem areas. For example, the procedure whereby files were requested (ie using a messenger service) appeared to cause unacceptable delays with access to information held in the Registry. There were also difficulties for staff who undertook priority work for such matters as public enquiries, and required access at times when the Registry was closed, eg at lunchtime and hours outside the normal working day. Staff in both departments found it necessary to maintain informal or social contacts to obtain information required for their tasks, and had some misgivings about the lateral flow of mutually relevant information.

Monitor

The technical staff had difficulty verifying the up-to-date status of information being used, which could be significant in terms of accuracy of work produced. Control of the movement of files (and journals, technical books, etc from the CSSU Library) depended on staff adhering to prescribed procedures, particularly those for registering file transfers between staff. These procedures were not strictly observed, resulting in temporary or permanent loss of items. In addition, it was felt that the problems were made worse by the files on loan being the only master copy; additionally, when files were held by staff for excessively long periods, a backlog of items for filing could accumulate .

Allocate Resources

A number of resource problems were also identified, relating to the lack of clearly defined responsibilities for microfilming, and the archiving/destruction of files. It also appeared that there were insufficient resources for maintaining localised filing systems in good order, considered to be a result of the departmental policy to centralise filing and associated staff.

Decide Criteria for Assessing Performance of System

It was suggested that the morale of Registry staff was low, and they appeared isolated from other parts of the departments with no appreciation of their role in terms of other functions, due to the lack of feedback about their performance in meeting the needs of the users.

14.2.5 Development of Functional Requirements Specification

In the latter stages of Activity 3, a requirements specification was developed, taking into account the problems and needs identified during the analysis. It was concluded that these could be addressed by two overlapping systems, described as:

1. A locating and accessing system (ie an index system), where references to information objects would be kept and cross-related to other items of information;

2. A storage system for maps, drawings and other technical documents which need to be updated frequently and are used in various branches of the Planning and Highways departments.

Functional Description of the Proposed Index System

The functional description contained a statement of the *purpose and main characteristics* of the index system, referring to the need to record, in a central index, all categories of files and documents generated, processed and received by the departments. It was also acknowledged that it would be necessary to determine which information objects should be indexed, accepting that certain brochures, circulars and so on would probably be excluded from this general requirement. The statement also recognised that, in terms of accuracy, integrity, consistency and up-to-date status, the quality of the index would need to be extremely high.

The *structure of the index items* was described in terms of the type of search facility required to support the work of branch and other clerks, for example, on the basis of title, keywords, or links to geographic map references. There was also a requirement to maintain consistency with past practice by generally following the file references of the departments which were then in use. An example was given of the type of index entry items that could be supported by the system:

- unique identification number for each file or document,

- title of file or document,

- short description of contents of item,

- primary reference code (= current references of Registry),

- originator of entry (person and their position in the organisation),

- authority for contents (person and their position in the organisation),

- home location of the item (eg Registry or branch filing system),

- current holder of item (when it has been loaned out),

- change history of information items (versions and dates),

- keyword list for use in topic searches.

The remainder of the functional description was concerned with elaborating on the ideas expressed by the conceptual model, explaining how they would be implemented in practice. For example, *collecting information* would be achieved by the Registry and branch clerks entering the information into the index on receipt, or after categorisation in terms of the topic area. Items generated internally would be recorded once they had been authorised for issue. *Accessing* the information would be improved by providing procedures that would be easier to use, and the structure of the index would enable a search on a variety of parameters, with each item adequately cross-referenced. *Maintaining* the index would be improved by establishing a specific role of system manager with responsibilities for the quality and integrity of the index, and for making exceptions to rules when required. Other management functions would be authorised to cover the deletion or amendment of certain fields. System *monitoring* would be enhanced by the collection of statistics (computer-aided) to influence archiving policy, the integration of local indexes with the central system, and, by demonstrating the benefits of the new system, encouraging greater adherence to required procedures.

Functional Description of Proposed Storage System

The concept underlying the proposed storage system was that files and documents would be subject to central management, achieved by keeping a record on the proposed indexing system. The functional description also recognised the need to agree and implement an archiving and purging policy, making use of the Council's central records office for storing information that had to be retained but was no longer in regular use.

The systemic requirements of the conceptual model were also considered. *Collecting information* would be improved by making the officer who generated or handled information objects responsible for transferring the master copy to the central storage system, a back-up copy being retained on file for use when the master is already on loan and further copies are required. *Accessing information* would be improved by showing the file location in the index, and making secondary copies available for issue if required.

Maintaining the storage system would be the clear responsibility of the CSSU, ensuring quality in a number of ways, eg by chasing up holders of master files, by inserting new objects into files, and by making secondary copies of files and documents. *Monitoring* would be based on the frequency of access to files, and the archiving policies of different types of information items.

14.2.6 The FAOR Technical Scenarios

During the final stages of the Case Study, two technical scenarios for implementing the requirements of the functional descriptions were developed. These can be summarised as:

1. A hybrid system consisting of a centralised computer index, and a variety of microforms used primarily to provide back-up copies of all information items.

Terminals to access the central index would be located in all branches, and staff changing the status of any document or files would up-date the index accordingly. A secondary copy of all information items generated or received within the departments would be held in the Registry on microfilm.

2. A fully computer-based system, where all objects (ie including maps, drawings and texts) would be filed electronically. This would be mainframe-driven with suitable terminals in all branches, and the person responsible for any information item would ensure that it was copied to the master file for other parties with the necessary authority to access.

The two technical scenarios are summarised in Figs 14.6 and 14.7. It was felt that these would allow a gradual transition from computer-supported indexing (with back-up copies of documents held on microfilm) to a fully computerised filing and retrieval system.

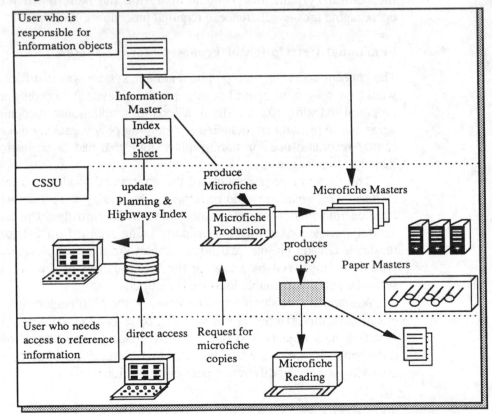

Fig 14.6 - Proposed Hybrid System

A detailed description of each technical scenario is contained in the FAOR publication, but certain points made by the project team are worth repeating here. It was noted at that time that there was a low level of information technology utilisation in the departments, consequently the cost of implementation would be a major factor when considering the design of either system.

The other significant points can be summarised as:

1. Both systems would require a critical mass of terminals to be effective.

2. The systems would only be effective if they were planned to converge towards an integrated system.

3. Electronic filing would require very large storage capacities for maps and technical drawings.

4. The filing and retrieval of graphic-based items would require interfacing to computer aided design workstations or digitisers, ie for entry, creation or updating of documents.

5. Considerable further work would be required to analyse and describe the information objects which would be candidates for indexing, and/or electronic filing.

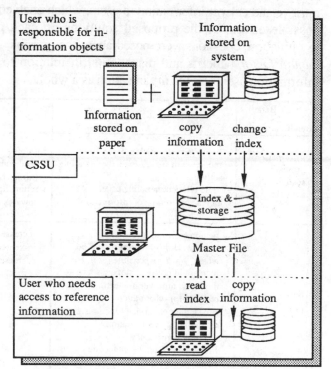

Fig 14.7 - Proposed Fully Computerised System

14.2.7 Activity 4 - Benefits Analysis

Having defined the functional requirements, and elaborated the technical scenarios for the proposed IPS, the benefit analysis instrument was used to investigate the possible improvements to the communication and cooperation processes between the providers and users of the service, as proposed during the method tailoring activity.

The main questions of the benefit analysis for this project were stated as:

- *How will the proposed changes affect the respective achievement of objectives of the information users and providers ?*

- *Which processes should primarily be supported to gain the most benefits for the IPS as a whole ?*

- *Which benefits (efficiency, effectiveness, human factors) will materialise in the implementation process ?*

Addressing and answering these questions was essentially a process of considering how existing work practices would be affected by each of the technical solutions, aided by a device known as a **measurement unit**, developed from the knowledge gained during the analysis activity. The measurement unit used in this study was similar to a flowchart showing the procedures undertaken to store, retrieve and deliver information to users, which enabled the analysts to consider the possible effects of the proposed solutions on each phase. In parallel with this examination, questions were posed about the possible *efficiency*, *effectiveness* and *human factor* benefits and disbenefits, in relation to the information providers, information users and the organisation as a whole.

Benefits / Beneficiary	+	-
Information provision system as a whole	- improved information bases - reduction of frustrating paper handling activities - more efficient handling of an increasing volume of information	- considerable investment in equipment training costs - need computer expert to manage system initial cost of putting data on system - setting up and regulating of policies towards archiving, media, storage location etc.
Service providers	- better control of information: know what information should be, where, who's responsible for it, etc. - decrease in physical storage space - improved turnover in meeting requests for information - broadening of service functions eg archiving, microfiching, ensuring consistency - increased personal contact due to new functions	- increase in manpower - increased number of stored information decrease in personal job satisfaction due to expected changes in work content
Service users	- easier access to relevant information - time savings in searching for and retrieving information, plus less interruptions in other work - better decisions made based on better-quality information - quicker and more effective response to enquiries from public- improved paper handling leading to more space at workplace	- specific responsibility for supporting administration of information - improved service will encourage more information searches - possible loss of personal contact with colleagues when searching for information

Fig 14.8 - Summary of Benefits Analysis

It should be noted that this exercise was not concerned with *cost-benefit* analysis, as at this point no potential costs had been derived or considered. Consequently the assessment was mainly a *qualitative* one carried out at a fairly high level, with little examination made of matters such as staff increases, equipment costs, technical feasibility, and so on. The results of the benefits analysis in this case are summarised in Fig 14. 8.

14.2.8 Conclusion of FAOR Study

Having completed the evaluation stage (Activity 4) and produced a broad statement of the requirements for an IPS, the part of the study using the FAOR methodology was completed. Design and implementation of the proposed system was outside the FAOR remit (Fig 14.9) but the proposals had raised the expectations of the two client departments, and it was decided that a follow-up study would be carried out using management services practitioners from the parent organisation, supported initially by an analyst from an external consultancy.

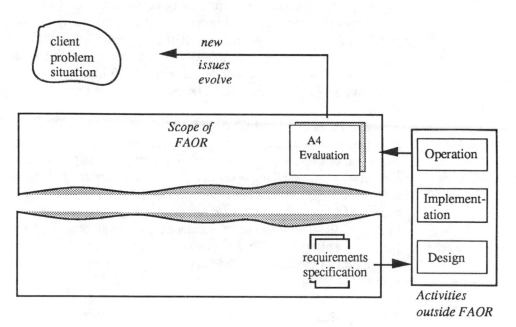

Fig 14.9 - Scope of FAOR Package

14.3 The Follow-up Review

On completion of the study, and a report to the client on the lines described above, the project team spent a number of months preparing the final documents for the ESPRIT Project as a whole, which formed the basis of the FAOR publication. It was accepted by the Highways and Planning departments that it was outside the scope of the methodology to progress the technical solutions to the point of implementation, which, in any event, could not have been undertaken by the FAOR team due to their disbandment after the final report had been rendered to the European Commission.

14.3.1 Second-phase Analysis Activity

In the early stages of the internal follow-up review, it was agreed that the proposed hybrid system (ie making use of microfilm supported by a computer index), rather than reduce duplication of documents, would tend to encourage it. By allowing a number of copies of the same document to be produced on request from the microfilmed back-up copy, branches would be likely to establish their own duplicate files. Furthermore, it was felt that the potential costs of providing the necessary microfilming equipment, plus additional staff time, would outweigh any benefits gained, particularly as this could only be regarded as an interim solution before progressing to greater computerisation. Accordingly, after the initial investigations, it was accepted by all the parties concerned that it would be inappropriate to pursue the microform aspects of the hybrid system.

However, at this point, the proposal that a computer index could be developed to improve the management of filing systems generally was considered worthwhile, and this became the focus of the early part of the study. It was also agreed that the options for storage/retrieval of information in the long-term would be examined, based on the FAOR proposals to move towards a fully computerised storage system.

Bearing in mind that the departments comprised a multitude of functions, and, at this point in the study, no detailed breakdown of these functions was available, two parallel exercises were considered necessary to progress the study, ie one concerned with clarifying *roles* and *information needs,* and the second exploring and quantifying the existing *information base,* as represented by the central and outposted filing systems. To *bridge the gap* between the outputs from the FAOR study and the development of a detailed user requirement, at this point systems ideas were merged with a more conventional fact-finding exercise. A brief description of the second phase activities and findings is given in the following paragraphs.

Investigation of Functions

To help the analysts understand the roles of the sub-groups of the departments, a conceptual model was constructed based on the findings of the Functional Analysis Instrument, ie that the main complementary roles of the two departments were:

- The production of the County infrastructure plan

- The design of detailed changes to the plan

- The control of design changes

- The execution of changes to the plan

- The maintenance of the infrastructure

In a manner similar to that described in earlier chapters, the model was expanded to reveal the lower-order activities (Fig 14.10), and each one cross-related to the branches and sections of the two departments.

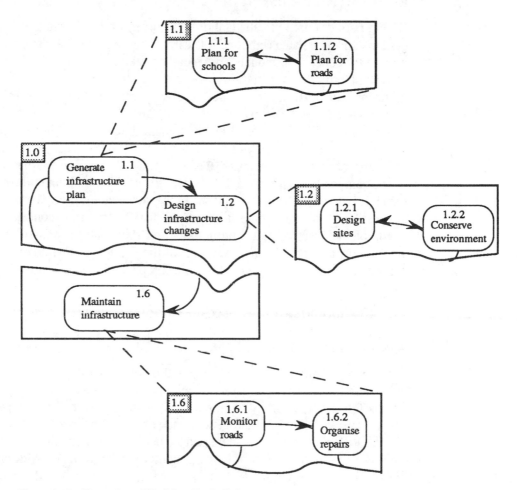

Fig 14.10 - Expansion of Model to Clarify Roles

This exercise provided a guide to the *logical* roles and activities of each group and the relationships between them, which, in conjunction with other background documents, was used as a basis for interviews with representative staff. These interviews followed a semi-structured format, each person being given a set of pre-interview notes posing questions about roles, responsibilities, information needs, and their usage of the existing filing systems. Although this appeared to repeat some of the work already undertaken by the FAOR team, because the analysts were now concerned with documenting the user requirements at a detailed level it was necessary to focus the questions more precisely and, taking account of the proposed technical solutions, to consider with the interviewees the potential advantages of information being indexed and stored by electronic means. As a result of an extensive series of interviews, a comprehensive functional description was prepared, and used to put the findings of the filing system examination into the proper context.

Examination of Filing Systems

Over a period of two months, a thorough survey of the central and outposted filing systems and associated activities was carried out to provide quantitative data related to the IPS conceptual model. This included taking a sample from each section of files to identify the type and quantity of file contents and the level of time-expired material, and also analysing daily activity rates for *file creation, up-dating, movements in and out, weeding/culling* and *file transfer or destruction*. The exercise proved to be extremely complex and time-consuming because of the sheer volume of material involved (ie in excess of 60,000 files), and the variety of different reference systems used. The examination not only confirmed the findings of the comparison stage of the FAOR study, but also highlighted the difficulties and expense that would be involved if the proposals were to be implemented at the level suggested; at the same time it helped to dispel certain misconceptions about the needs of the users overall. The output from this analysis was summarised in the final report as a functional description of the filing systems, and used as the basis for the detailed user requirement that was prepared for the client.

14.3.2 Findings of the Second Phase

Review of Functions

Eventually a full picture emerged of the roles of the functional groups, their information needs, and the structure and use of the existing filing systems. At first it was felt that the variety of reference systems in use had developed because of inept control mechanisms and general lack of discipline in applying procedures, but as the analysts' knowledge increased, it became apparent that the references had evolved to suit the needs of users, needs which were considerably different throughout the departments. For example, all sections that dealt with Planning Applications from District Councils preferred to use the District Council reference as a common base for their filing systems. In other instances, specific case files were raised to bring together all information related to individual projects, which could have a life-cycle of several years; some sections were primarily concerned with specialist subjects on a geographical basis (eg all the archeological sites in each district of the County); certain support services required files on matters of policy, or personnel, or finance, and so on. Overall, with the possible exception of Planning Applications, most references were *subject-based*, and, although in broad terms certain subject matters were the same, at a working level the information needs were totally dissimilar, reflected in the manner of structuring the reference codes. The detailed examination of the organisation of these references confirmed the view that there was little or no commonality between the systems used, and within each system, reference fields were utilised for any combination of alphanumeric codes that suited the purpose of the prime users. (At one point it was commented that *"There are an infinite number of ways of organising our file references, and most of them have been used already!"*)

It was also obvious that there was no common set of *criteria of needs* (ie in terms of access times, location, form, archiving etc) that applied to all groups, a conclusion that had serious implications for the proposed systems. For example, some functions required fast and frequent access to both active and dormant files; others managed efficiently with slower access times, but had a need for more extensive cross-referencing; some items had a limited life only, whereas information relating to policy and major structural changes had to be kept indefinitely. It did not appear appropriate, therefore, to develop one global user requirement for the two departments as implied by the FAOR proposals, and, in any event, this would not have been possible without extensive reorganisation of the filing and reference systems as a whole.

Review of Index Requirements

During the second phase, the analysts had also investigated systems that were then available for filing system support, using statements given in the FAOR report as a basis for defining the type and level of system required, eg:

"All major categories of file and document, either generated and processed by the branches of H & P, or received subject to later referral as part of the highways and planning functions, should be recorded (for subsequent reference) in a central index."

and:

"Many of the fields proposed for the index item will support different kinds of search, e.g. on the basis of title, keywords, last date of updating, or linked to geographic map references."

These and other statements indicated that a comprehensive computerised *records management* system was intended by the proposals, ie one that required all activities and events associated with the files, and the documents they contain, to be recorded. Whereas the FAOR team took a high-level global view of the departments and the CSSU, in the second phase all the filing systems were examined and quantified in fine detail, and existing problems and long-term needs were discussed with filing and branch clerks, and the potential customers of the system. This detailed investigation revealed that, although in broad terms the earlier proposals were supported, the total number of files held by the departments (ie approximately 60,000), and the variety of ways in which outposted files were managed and used, would make the cost of providing a sophisticated computer system throughout excessively high, compared to the benefits that could accrue.

The analysts were unable to identify any software available at that time that could have supported this volume and variety of files without considerable development. Based on an earlier estimate of £70,000 for computerised support of approximately 20,000 files in another department (without networking costs), a similar system for the Highways and Planning departments, allowing for some economy of scale, could have cost in excess of £150,000 at the prices that were then current.

In addition to the obvious costs of hardware/software, greater disciplines would have to be imposed on all users of the filing systems in order to keep the computer record up to date; for example, all movements of files in and out from their home base would have to be logged by the user, including branch officers who have direct access to outposted files. It was also likely that all incoming correspondence would have to be recorded before circulation in order to maintain system integrity, which would be difficult to control as some correspondence was passed directly to branch officers without being opened in the Registry. The involvement of technical staff was recognised by the the FAOR team in the functional descriptions of the proposed systems; however, the cost of this involvement had to be considered carefully. Although these actions could eventually lead to improved records management generally, the high cost in terms of staff involvement that would result was felt to outweigh the benefits that would be gained.

Long-term Storage of Information

The FAOR proposals for storage of information in the long term using a fully computerised system were also investigated thoroughly, particularly as these had a bearing on the manner in which the index system would be developed. Despite the earlier reservations about the differing user needs, it was felt that, because both departments were dependent on *images* in their operations, eg plans, drawings etc, rather than just textual documents, there could be considerable potential for the use of **Document Image Processing (DIP)** systems. Accordingly, an investigation of the current state of the DIP market was carried out and the results presented to the client. This investigation indicated that, although the potential existed for use of such systems by the departments, at that time the costs were high and there was a general lack of standards and a limited amount of expertise available, and it was felt to be unwise to consider any major capital investment. Once again, because the needs of each group in the departments varied considerably it would not have been cost-effective to consider a general global use of DIP systems. However, because of the immense potential for the integrated (ie raster/ASCII/vector) systems that were under development, the point was made that the strategy could be to plan their eventual use for those functions where they could be of value. In particular, the purchase of a stand-alone system for specified functions was considered beneficial, as a pilot project to encourage the learning process for subsequent extended use.

14.3.3 Conclusions

That the FAOR team had identified a significant problem area was indisputable; however, the extent that the proposed solutions could be attained, given the related costs and other demands on available finance, was questionable, and it was necessary to review the solutions in light of the more detailed analysis. The findings from the investigation, as given above, caused both the client and the analysts to review the situation and concentrate first on the general organisation of the filing system as a whole, to determine if it were possible to achieve some degree of commonality as a prerequisite to the development of a computer support system. Consequently, a Filing Working Party was established and, after extensive examination of the reference construction, it was found that the only changes considered viable (ie that could be made without having to completely revise the titles of some 60,000 active files) were to the first three fields of the reference structure, establishing the basis for a limited computerised search facility in the proposed index system, rather than the extensive facilities envisaged by the FAOR Team. Additionally, to address the problem of large amounts of time-expired material being held in office space, a positive weeding exercise (based on an identifying dormant files by the date they were last used) was carried out, also reducing the number of file titles that would need to be held on the computerised index.

14.4 Development of a User Requirement

In summary, it was concluded from the initial examination of the FAOR proposals that the development of a global user requirement for the introduction of either computerised indexing and storage systems was not appropriate, or possible, at that time, and it was necessary to take an evolutionary bottom-up approach that recognised not only the limitations and costs of the current market in software and equipment, but also the particular needs of individual functions. In order to develop a suitable strategy to take these factors into account, a further examination was carried out to determine :

a. The extent that computerised indexes/records management systems could be of value to *individual* functions within the two departments.

b. Which functions would benefit by the introduction of a DIP system, either as a stand- alone system or as part of an eventual network.

These examinations progressed from the work already undertaken, and further discussion with staff of both departments. It was then necessary to prepare a strategy document showing how the eventual development of a comprehensive system could be achieved, and indicating where pilot schemes would be introduced for both filing support and the use of DIP. At this point, it became apparent that the systems being considered had lower priorities than other current developments, and as a result, the proposed pilot studies for DIP systems were discontinued, and the analysts concentrated on the development of the index system only. The strategy finally agreed consisted of a number of actions to be taken progressively, ie:

1. Implement the changes to the filing references proposed by the Filing Working Party, to achieve some degree of commonality.

2. Following the production of a suitable user specification, introduce limited computer support of the central filing system to assist with records management and indexing.

3. Review the requirement for computer support of outposted files as an added value factor once a department network had been installed (Note: this was due to take place after the departments had moved to the new accommodation).

On the basis of this strategy, the analysts produced a user requirement for the index system; however, this did not compare with the idealised IPS that was originally anticipated as a result of the two extensive studies. Nonetheless, it did seek to address the problems identified during the FAOR analysis, albeit at a lower-level of attainment in the initial stages. For example, amongst other things, the specification aimed to achieve:

● Significant reductions in the time required to retrieve files

● Immediate cross-referencing to related files, using word-search facilities

● Improved knowledge of file whereabouts once issued

● The elimination of unnecessary duplication of files

● An analysis of file events, as the basis for improving records management procedures

It was also felt that the system proposed in the final user requirement would lead to improvements in the working conditions in the Registry, and, by introducing more interesting and rewarding procedures at this level, generally improve staff morale. Management of the the filing system would be aided by the introduction of the disciplined procedures required to operate the computer system, and by an improved awareness of the overall state of the system at any time. The eventual extension of the index to include outposted files would bring them under closer scrutiny and control, and, although a degree of consistency would be achieved by revising the file references to the level recommended by the Working Party, branch officers would still retain considerable discretion to organise their files in a manner that suited their particular needs. The system would be configured so that the file database could be interrogated against a wide variety of search parameters (eg officer name, file reference, file titles, etc), and the basic structure of the index entries reflected that given in the FAOR functional descriptions.

At the time the analysts and the client were somewhat concerned that it was not practicable to implement fully either of the FAOR technical solutions, as expectations had been raised about the final outcome before the implications of these solutions were known in any detail. However, taken overall the exercise had focussed attention on the problems associated with the filing systems of the two departments, and many of them had been alleviated by the time the study was completed, albeit using manual methods rather than technical solutions. There was also a reasonable and sound expectation that, on the basis of the initial computer support installation, many of the other problems identified by the FAOR team would eventually be addressed and solved. It is worth recalling some of the advice given earlier about soft systems studies, ie that the analyst should not attempt to convert a *problem situation* to a *situation without problems,* but consider how to improve it in a manner and to the degree that is attainable, given the influencing factors in the situation at the time of the study.

14.5 Conclusions for the Systems Approach

The SSM is only one part of the total FAOR package, and as such was used to good effect to identify and select a specific problem area in the client organisation, and to complement the other analytical instruments. It also provided a basis for the subsequent investigation, encouraging the analysts to approach the second phase in a structured and methodical manner. The idealised FAOR solutions, although impractical to implement at the time, nonetheless provided a goal that could be pursued via a strategy that would recognise the speed of technological developments, the availability of finance, and other priorities of the two departments. It was always accepted that the FAOR study was essentially a research project, and any reservations about the final outcome in relation to the extensive analysis that was undertaken must be viewed in this context. However, in terms of the use of soft system analysis for everyday project work, there are some useful lessons to be learned from the FAOR experience.

A parallel can be drawn with any examination that uses SSM to draw out ideas about improvements before using more conventional techniques to develop those ideas to a satisfactory conclusion. It is important that the potential costs of *actions to improve,* in terms of extended research, development, equipment and so on, are also considered when debating possible changes with the client, together with an examination of conflicting priorities for whatever finance and other resources are available. The danger otherwise is the development of utopian ideas about moving towards ideal systems without true recognition of the problems involved, ideas that, at a global level of consideration, can also seem appealing to the client. In addition, it can be demotivating for staff who are subjected to a prolonged investigation that does not result in tangible benefits that offset the level of disturbance caused.

The soft systems approach often leads to conclusions about fundamental problems in an organisation, problems which in turn would require fundamental changes before they could be resolved. From these conclusions it may be possible to derive a *high-level* strategy for change. If, however, the expectations of the client are of a lower order, the analyst must be prepared to use the idealised solutions simply as a knowledge base to determine where *acceptable and attainable* improvements can be made. The exercise of decomposing the system models will aid this process, progressively bringing the systems ideas closer to what can be observed in practice; imaginative use of unsophisticated computer software can also help, particularly where the level of detail makes manual analysis impracticable; simple *matrices* similar to those employed to good effect in many of the studies described in this book can also be of benefit; not forgetting, of course, the significant value of discussing ideas with experienced staff from the organisation being studied. Other analysts may well develop different ways of making the connection; however, no matter how close the systems ideas come to the real-world activities, it will always be necessary to temper those ideas with a firm grasp of the practical effects of any changes.

The follow-up work carried out after the FAOR study involved an extensive *fact-finding* exercise, where a number of parallel investigations of real-world activity were undertaken to consider more precisely the effects of the proposed changes, in terms of data quantities, costs, staffing implications, and technical feasibility. Having completed these, it was then possible to develop a strategy for implementing solutions that were *feasible* and *desirable* to implement in the existing circumstances. The main lesson to be learned from this and certain other studies where soft systems thinking has been used, is that analysts, when making specific proposals for change, must take account of all the factors that determine the acceptability of such changes; in this way there is a better chance that the so-called pragmademic gap will be successfully bridged (Fig 14.11).

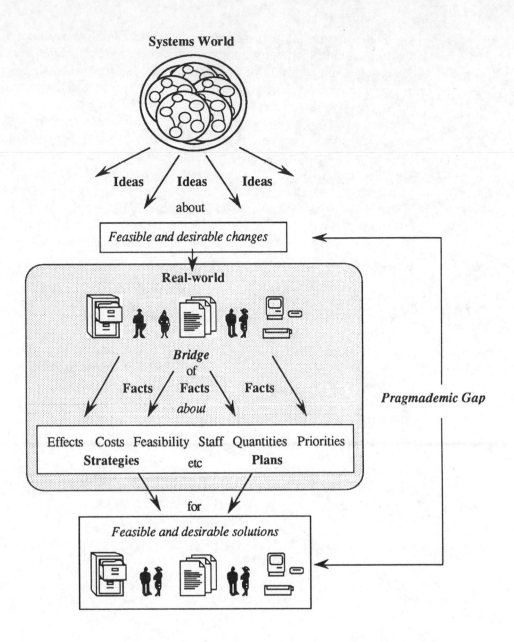

Fig 14.11 - Bridging the Pragmademic Gap

15 Application Guidelines

15.1 Introduction

Having entered the soft systems world as an untrained novice some years ago, and endeavoured to apply the theories to problem situations within my normal sphere of activity, aided only by the few textbooks that were then on the market and occasional contact with other exponents of the approach, I am keenly aware of the difficulties that some practitioners may have, not only with understanding the principles but also with putting them into practice. I hope that this book overall will provide additional assistance in both respects, particularly for those who may not have had the advantage of formal systems training. In the preceding chapters there is a lot of advice about how to apply systems ideas, but, to encourage a more complete understanding, it has been necessary to surround this advice with detailed explanations and examples. Consequently, the reader may have reached this point in the book without a firm idea of how to start, or progress, a project that makes use of soft systems thinking. Throughout the book I have tried to make it clear that the approach is not prescriptive, and successful applications depend a great deal on the intellectual and creative skills of the analyst. Nonetheless, it is always useful to have some general guidelines, and in this final chapter I have gathered together many of the points that, from my personal experience, are worth bearing in mind when embarking on such a project. As detailed explanations are provided earlier in the book they are not given here, but I have included appropriate chapter references in the text.

15.2 Applying the SSM

An overall summary of the SSM is given in Chapter 4, which includes a valuable statement of the outputs from the main stages in conjunction with the constitutive rules that apply. In addition, there are a number of useful devices and hints that can help whenever the SSM is used, and these, together within certain pitfalls to be avoided, are summarised in the following sections.

15.2.1 Getting the Picture Together

Chapter 5 refers to the process of exploring a situation and expressing the main factors as a rich picture. As a guide to construction, these pictures should reflect relevant *structures* and *processes,* and give an overview of the *climate* of the particular situation being addressed, together with any *issues* that are pertinent at the time of the analysis. In practice, it isn't necessary to concentrate on including these factors, as they will generally appear in some form in the finished product. From the outset of a study the picture can be developed simply by:

- Noting down (in picture form) any points about the main characters, policies, locations, finance, constraints etc that seem relevant when reading through background papers, correspondence and other related material.

- As interviews and other meetings are held, adding the issues to the picture as they arise, making sure that the source of these can identified, or the reasoning behind the selection of an issue can be substantiated. Bear in mind that issues should reflect **fundamental** concerns (in relation to the level of study), and might also include those problems that arise **frequently**, or are *in vogue* at the time of the study.

- Using the *interview analysis* technique where there is a need to draw conclusions about issues from a large number of interviews notes, discussion papers etc. This analysis can be carried out manually by examining related complaints, grumbles etc and assessing the root cause of the expressed problems, or by utilising the facilities of a computer database to record, assess and regroup a series of minor points.

Do not be overly concerned with the *correctness* of rich picture, particularly in terms of format and layout, although the conventions used should be understood by all the analysts involved. The term 'rich' implies that the pictures should contain a wealth of information relevant to the study; however, trying to include too much detail can result in them becoming extremely cluttered, and it is often necessary to summarise the detail (using the interview analysis technique if required) and redraw them as the study progresses. Two further points are worth remembering ie:

- When using the pictures as the basis for discussion with the client or representatives, be conscious of the sensitivity of any comments or issues that they reflect.

- The clarity and definition of rich pictures can be aided by using computer graphics; however, attempting to construct them on a computer from the outset can impose artificial constraints, and the original pictures should always be drawn freehand.

15.2.2 Developing System Models

Apart from the value of the pictures in terms of summarising the main factors about a situation, they are used as the basis for selecting relevant systems when preparing root definitions and associated conceptual models. This is carried out by first defining what the system *is* that is being modelled, using the CATWOE mnemonic as a guide (Chapter 6).

Selecting a Viewpoint

The most important part of the CATWOE analysis is the determination of the **Weltanschauung** or **Worldview** and the associated **Transformation**, which normally precedes consideration of the other CATWOE components. If the concept of Weltanschauung is difficult to grasp, consider instead the different *viewpoints* that are relevant to the situation being examined. As a starting point, study the rich picture and *make a list* of the main characters or groups of people that could have an interest (particularly those who are influential in some way) and then consider the systems that can be perceived from their points of view. For instance, the staff, management, shareholders, customers etc could each view the enterprise in different ways, and the list will encourage the analysts to focus on each perspective in turn, leading to ideas about the systems that could be modelled. In most cases it is also useful to develop a statement of the *primary task* of the enterprise, based on a neutral assessment of what is actually happening or is being achieved (such as *making motor cars,* or *providing a specified service* and so on).

The development of a number of root definitions and associated models will be enlightening; however, if time is a constraint then it may be necessary to be more selective and pursue only the viewpoints that are *most* relevant, such as those of the main client or other influential persons, ie people whose agreement may be needed before changes can be implemented.

Transformations

Once a viewpoint has been selected, the subsequent CATWOE analysis is made easier by first considering the transformation that could be taking place in the selected system, using as a starting point a simple sketch of the conversion process, ie *input-transformation-output* (Fig 15.1).

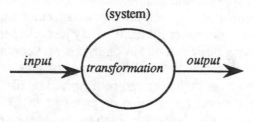

Fig 15.1 - Sketch of Conversion Process

Bear in mind that:

- It sometimes helps to decide first the desired output from the system and work backwards to consider the input, leading to a statement of the desired transformation.

- Inputs and outputs should be at the same level of abstraction; in other words conversion of an abstract notion (such as *customers needs)* to a tangible form (such as the *specific* goods or services that are supplied*)* is not supportable in a root definition. It is, however, acceptable to illustrate in the conceptual model secondary inputs/outputs in support of the main transformation, and some of these (such as resources, information etc) could have a tangible form in the real situation.

Environment and Wider Systems

When deciding the environment of the system being modelled, keep in mind the simple definition that applies in this context, ie it *influences but doesn't control* the system, and also the notion that the environment provides the inputs to the system and receives the outputs from it. Don't confuse the environment with the *wider systems of interest,* which, although providing inputs and receiving outputs, exercise *control* over the system being considered, normally by setting the standards, rules and so on; additionally, the *wider system* often provides the resources that are used by the subject system.

To avoid delaying the progress of a study by being too concerned with defining the environment or other CATWOE elements correctly, it may be necessary to merge some real-world facts with the systems thinking, for example by considering the relationships between the group being studied and other parts of the organisation, and setting the system boundary accordingly. This might appear contrary to the idea of keeping systems thinking and real-world aspects separate, but it can provide a compromise that ensures that progress can be made; however, such compromises should be recognised as such and not made as a matter of course.

Other Factors

Similarly, because system boundaries are not synonymous with functional or organisational boundaries, difficulties can be encountered when defining the *actors, owner(s)* and *customers* of a Human Activity System. During the initial development of root definitions these factors should be considered in *systems terms only,* otherwise the analyst may just relate them to the subject organisation , overlooking other individuals or groups who contribute to the system as defined. If progress is unduly delayed by failing to agree on sufficiently 'pure' definitions, then the systems thinking can be tempered by using such devices as the *hierarchy of control* model (as covered on pages 81 to 83 of Chapter 6). Don't forget that *customers* can be defined for each lower-order element of the model, as well as for the system as a whole.

Expressing the Root Definition

As a final point about the root definition, it is sometimes useful to express it as a statement to see if it makes reasonable sense, and to double-check that all the CATWOE factors have been considered. There is no precise form for such a statement, but the following structure has been found useful in practice:

'A system owned by *(Owner as defined)*, operated by *(Actors as defined)* to carry out/do/provide etc *(ie the Transformation as defined)* for the *(Customers as defined)* in an *(Environment as defined)*'

It is not necessary to include the viewpoint taken in a statement of the root definition, as this will have been explicitly considered in the early part of the CATWOE analysis. One further point is worth remembering; the root definition is concerned with **what** is being achieved, not **how** it is done, and any inclusion of a 'how' statement should be shown as a constraint, eg improving productivity *by the use of incentive bonus schemes*.

15.2.3 Constructing the Conceptual Model

Model styles can vary considerably, but there are certain general points to be conscious of when constructing them, ie:

- The components of the model are expressed as verbs.

- If the model contains too many sub-systems/activities, it is likely that there is a mix of resolution levels. Five to ten sub-systems is accepted as the normal range, which can then be decomposed as required to show lower-order activities.

- It is often difficult to show on the model precisely how the components interact (eg by a transfer of information or resources), particularly when drawing high-level models where such interactions may be extremely complex. However, it is important to show the links between components, even if the detail is not included.

- The 'monitoring and control' sub-system, which relates to all components of the model, can be shown as a separate element to avoid cluttering the illustration.

- The process of expanding or decomposing the model should be carried out by developing a root definition for each sub-system, taking into account that these are subject in turn to 'monitoring and control' at the new level of definition.

- The *formal systems model* provides an invaluable checklist for ensuring that all the general system characteristics have been considered when developing Human Activity System models.

- The ideas generated (ie about resources, interaction, measures of performance, decision-making processes etc) when checking with the formal systems model can also be used during the comparison stage of the SSM to determine if these factors are represented in the actual situation.

15.2.4 Exploring the Real Situation

Chapter 7 examines in detail the methods for making the comparison between the conceptual models and the real situation. The main points to remember are:

- System models are seldom synonymous with organisational structures or functional groupings, and high-level models reflect **what** is being achieved rather than the manner of achieving it (ie **how** this is done). Expanding the model will, however, progressively reveal activities that *may* be observable in practice.

- There are a number of tried and tested approaches for making the comparison (eg *general discussion and observation, question generation, model overlay*), but these are not intended to be exhaustive; analysts should endeavour to make imaginative use of the ideas generated during the systems thinking stages to examine what is happening in practice.

- Expansion of the model (ie *extended analysis)* can be aided by using a computer database to record and retrieve each activity, associated resources, information, and measures of performance.

- To maintain links between the main model and the outputs of the expansion, an 'audit trail' should be established by numbering each component of the model in turn, then extending the numbers on a decimal basis as the expansion progresses. In addition, as discussed in Chapter 11, the use of a 'hierarchical scalar' can assist with obtaining an overview of the main components of an expanded model (Fig 15.2).

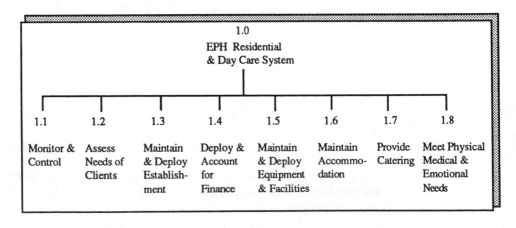

Fig 15.2 - Example of Hierarchical Scalar (from Chapter 11)

15.2.5 Agreeing and Implementing Changes

As explained in Chapter 8, the output from the comparison stage will be an *agenda for debate* about areas where some improvement appears to be possible; in practice this may comprise a list of problems that the analyst has identified for discussion with the client so that one or more can be selected for further study. The changes that will be deliberated at this point are those that are considered *desirable* in systems terms, and culturally *feasible* in light of the prevailing attitudes within the organisation. In addition, experience has shown that it is equally important to make a provisional assessment of factors such as development and staffing costs, and the technical feasibility of pursuing any proposed computer initiatives.

Whether or not a formal debate is held will depend on the purpose of the examination and a number of other factors, but it is obviously necessary to seek the client's opinion before deciding the direction that a study should take once a number of problems have been identified. Discussions should also be held when relevant systems are being considered during the early stages, and when some specialist input (ie from the organisation) would be advantageous to the modelling exercises.

15.3 Informal Applications

Chapter 9 covers a variety of different situations where systems ideas can be used informally to help with general analysis work, and these can be summarised as:

- Making a sketch in rich picture form when gathering information for any type of study, or when attempting to obtain an overview of a complex situation.

- Without deriving a full root definition, modelling the activities required to undertake a particular process such as project planning, performing a task or set of tasks, developing ideas about new functions, and so on.

In summary, systems concepts and principles can, with a little imagination and effort, be utilised to good effect in virtually any situation, including those where computer systems development is taking place, either in the limited sense described above, or by making use of the full SSM package to put such developments into an organisational and social context.

15.4 Final Comments

In the book I have attempted to give a practical interpretation of the soft systems approach, an interpretation that has been found to be of value in many studies with which I have been associated. To achieve this, and also to attract the interest of would-be exponents of the approach who have not had the opportunity to undergo an extensive systems education, it has been necessary at times to over-simplify the explanations of concepts and applications, and condense them from those developed by system researchers over many decades. Any reader who wishes to pursue the concepts further and obtain a more complete understanding should first explore the main textbooks on the subject, ie by Peter Checkland and Brian Wilson, and then delve into the many others that examine systems ideas, some of which are given in the list of references.

On the whole, I hope that there is sufficient material in the book for analysts of any persuasion to make a start on the process of using system ideas to investigate and understand human activity; at the same time, systems students and researchers may well benefit by studying some of the interpretations and practical examples given here, reflecting as they do a *Weltanschauung* with which they may not be familiar.

In broad terms, because soft systems thinking represents a unique way of encouraging an observer to learn about a situation, it should appeal to virtually anyone who has an enquiring mind, a notion reflected in the closing paragraphs of *Systems Thinking, Systems Practice;* as I have made frequent reference to this work throughout, it seems appropriate to finish my book with the following quote from Professor Peter Checkland, who says about soft systems thinking:

"It will appeal to all those people in any discipline who are knowledgeable enough to know that there is much they do not know, and that learning and re-learning is worthwhile. For such people a systems approach is not a bad idea"

Which is not a bad idea either !

References and Further Reading

Ashby, W R (1956) *An Introduction to Cybernetics*, Chapman and Hall

Beer, S (1976) *Decision and Control*, John Wiley and Sons

Beishon, J (1980) *Systems Organisations: The Management of Complexity*, Open University Press

Beishon, J et al (1974) *Systems, Organisations and Management*, Open University Press

Bentley, T J et al (1984) *The Management Services Handbook*, Holt, Rinehart and Winston Ltd

British Standards Institution (1979) *Glossary of Terms Used in Work Study and Organisation and Methods (O&M)*, BS 3138

Burns, T and Stalker, G M (1966) *The Management of Innovation*, Tavistock Publications

Checkland, P (1981) *Systems Thinking, Systems Practice*, John Wiley and Sons

Checkland, P (1972) 'Towards a Systems-based Methodology for Real-world Problem Solving', *Journal of Systems Engineering* (2, pp 9-38)

Checkland, P and Wilson, B (1980) 'Primary Task and Issue-based Root Definitions in System Studies', *Journal of Applied Systems Analysis* (7, pp 51-54)

Checkland, P B (1978) 'The Origins and Nature of 'Hard' Systems Thinking', *Journal of Applied Systems Analysis* (5, pp 99-100)

Churchman, C W et al (1957) *Introduction to Operations Research*, Wiley; Chapman and Hall

Churchman, C W (1968) *The Systems Approach*, Dell

Clemson, B (1984) *Cybernetics: A New Management Tool (Cybernetics and Systems Series)*, Abacus Press

Drucker, P F (1979) *Management*, Pan Books

Emery, F E & Trist, E L (1972) *Towards a Social Ecology, Contextual Appreciation of the Future in the Present*, Plenum Press

Emery, F E et al (1969) *Systems Thinking, Selected Readings*, Penguin

Flood, R and Carson E, R (1988) *Dealing With Complexity: An Introduction to the Theory and Application of Systems Science*, Plenum Press

Gane, C and Sarson, T (1979) *Structured Systems Analysis*, Prentice Hall

Handy, C B (1981) *Understanding Organisations*, Penguin

Hertzberg, F (1966) *Work and Nature of Man*, Staples Press

Hirscheim, R (1985) *Office Automation: A Social and Organisational Perspective*, John Wiley and Sons

Jackson, M C (1982) 'The nature of "soft" systems thinking: The Work of Churchman, Ackoff and Checkland', *Journal of Applied Systems Analysis* (9, pp 17-29)

Katz, D and Kahn, R L (1966) *The Social Psychology of Organisations*, J Wiley and Sons

Lawrence, P R and Lorsch, J W (1967) *Organisations and Environment, Managing Differentiation and Integration*, Havard University Press

Lawrence, P R & Lorsch, J W (1969) *Developing Organisations, Diagnosis and Action*, Addison and Wesley

Maslow, A H (1954) *Motivation and Personality*, Harper

Mayo, E (1948) *Hawthorne and the Western Electric Company; The Social Problems of an Industrial Civilisation*, Routledge and ?

Mcloughlin, P (1986) 'Soft Systems Methodology - its role in management services, or finding a nut to fit the spanner', *Journal of the Institute of Management Services* (30, pp 16-20)

Mcloughlin, P & Brown, G (1989) 'Procedure Audit', *Journal of the Institute of Management Services*, (33, pp 12-15)

Mintzberg, H (1979) *Structuring of Organisations: a Synthesis of Research*, Prentice Hall

Mumford, E and Henshall, D (1983) *Designing Participatively, A Participative Approach to Computer Systems Design*, Manchester Business School

Nadler, G (1970) *Work Design - A Systems Concept*, Irwin and Dorsey

Naughton, J (1977) *The Checkland Methodology: A Readers Guide*, Open University Press

Newell, A and Simon, H (1972) *Human Problem Solving*, Prentice Hall

Patching, D C (1987) 'Soft Systems Methodology and Information Technology' *Journal of The Institute of Management Services* (30, pp 16-19)

Patching, D C (1989) 'Understanding Rubik Situations', *Journal of The Institute of Management Services* (33, pp 18-21)

Petri, C A (1980) *Introduction to General Net Theory*, Brauer

Pugh, D (1984) *Organisation Theory*, Penguin

Pugh, D S et al (1987) *Writers on Organisations - 3rd edition*, Penguin

Raybould, E B and Minter, A L (1976) *Problem Solving for Management*, Institute of Management Services

Rice, A K (1958) *Productivity & Social Organisations:the Ahmedabad Experiment*, Tavistock Publications

Rice, A K (1963) *The Enterprise and Its Environment*, Tavistock Publications

Schafer, G et al (1988) *Functional Analysis of Office Requirements: A Multiperspective Approach*, John Wiley and Sons

Simon, H (1976) *Administrative Behaviour. A Study of Decision-Making Process in Administrative Organisations - 3rd edition*, Free Press

Smith, B L R (1966) *The RAND Corporation:Case Study of a Non-profit Advisory Corporation*, Harvard University Press

Sofer, C (1972) *Organisations in Theory and Practice*, Heinemann Educational Books

Taylor, F W (1911) *The Principles of Scientific Management*, Harper and Row

Trist, E L and Bamforth, K W (1951) 'Some social and psychological consequences of the longwall method of coal getting', *Human Relations* (4, pp 3-38)

Whitmore, D A (1975) *Work Measurement*, Heinemann

Wilson, B (1984) *Systems:Concepts, Methodologies, and Applications*, John Wiley and Sons

Wood-Harper, A T and Antill, L and Avison, D E (1985) *Information Systems Definition: The Multiview Approach*, Blackwell Scientific Publications

Woodward, J (1980) *Industrial Organisations, Theory and Practice - 2nd edition*, Open University Press

Glossary of Terms

Activity

A sub-set of actions considered necessary to achieve a desired transformation in a Human Activity System.

Actor

A person who undertakes one or more of the activities of a Human Activity System.

Agenda for debate

A summary (for discussion with a client) of problems and areas of potential improvement that have been revealed when comparing conceptual models with the actual situation.

Analyst

Used in the context of this book to refer to *"all those who observe a situation with a view to improving it"*.

Audit criteria

Performance measures selected to assess the effectiveness of a procedure when using the *Procedure Audit* method.

Audit trail

A means of relating the by-products of extended analysis to the main system model, normally making use of a decimal coding arrangement.

Basic systems template

A diagram showing prominent characteristics of systems (ie inputs, transformation, sub-systems and outputs), used as an aide memoire when considering situations in systems terms.

Boundary

In a systems model, the area that the *regulating* or *decision-making* mechanism is able, or authorised, to control.

CATWOE

A mnemonic made up of the initial letters of the factors that are included in a well-formulated root definition (ie *Clients/Customers, Actors, Transformation, Owner, Weltanschauung, Environment*).

Client

The person or persons who set the directive for a study, and to whom the subsequent findings are reported.

Clients/customers

In the CATWOE mnemonic, those who benefit from, or who are affected by, the outputs from a Human Activity System.

Climate

Defined as the *relationship between structure and process* in a rich picture; in effect, the overall impression that the picture gives of the prevailing circumstances.

Closed systems

Systems that are self-contained and do not interact with the *environment.*

Comparison stage

In the Soft Systems Methodology, the stage where the real situation is examined in light of the models developed during the *systems thinking* phases.

Conceptual model

Defined as *"a systemic account of a Human Activity System, built on the basis of that system's root definition, usually in the form of a structured set of verbs in the imperative mood"*. In effect, a diagrammatic description showing an ordered arrangement of the sub-systems or activities considered necessary to accomplish a desired transformation, using verbs to describe the system components.

Constitutive rules (for the SSM)

The basic rules that appply to the Soft Systems Methodology and distinguish it from other approaches.

Continuity

In systems terms, the expectation of continuing existence or stability, and the ability of a system to recover from disturbances.

Control

Formally defined as *"the means by which a whole entity retains its identity and/or performance under changing circumstances"*. Exercised in a Human Activity System by the decision-making or regulating mechanism.

Cultural Feasibility

A criterion for judging the acceptability of changes in light of an organisation's specific norms and values.

Decomposing (conceptual models)

The process of expanding a system model to reveal lower-order sub-systems and activities.

Designed/man-made system

A set of interacting components or devices that are engineered or manufactured to operate as a whole entity; also taken to include art forms contrived and constructed by man.

Emergent properties

The new properties displayed by a whole entity that are a result of the interaction of the component parts, and which would not exist if the parts were separated.

Environment

That which exists outside the boundary of a system; defined as *influencing but not controlling* the system, and taken to provide *inputs* to, and receive *outputs* from, the system.

ESPRIT

European Programme for Research into Information Technology; a programme sponsored by the European Economic Community to promote research into information technology at the pre-competitive stage of development.

Expression stage (of the SSM)

The stage of the Soft Systems Methodology when all the main factors affecting a situation are displayed in graphical form as a rich picture.

Extended analysis

The process of expanding a conceptual model to identify lower-order sub-systems, activities, information needs, resources and measures of performance. Can be facilitated by the use of a computer database to record and retrieve the by-products of the analysis.

FAOR methodology

A framework for undertaking a *Functional Analysis of Office Requirements*, developed as part of the ESPRIT programme.

Feasible/desirable changes

The changes to an organisation that are considered acceptable in *cultural* and *systems* terms.

Feedback

Formally defined as *"the transfer of part of the output back to the input"*; in effect, a message in some form that indicates compliance with, or a deviation from, preset parameters, signifying when control action is required in a system.

Formal systems model

A general systems model listing the factors that characterise a Human Activity System.

Functional groupings

In an organisation, a formal arrangement of persons with similar skills or duties.

Hard systems analysis

The prescriptive use of techniques to address clearly-defined problems, especially those concerned with computer developments.

Hard/soft division

The distinction between methods of examination that address clearly defined problems (ie those that are suitable for the application of prescriptive *techniques*), and others that are used when the problem is not clear at the outset, and a preliminary investigation is required to *identify* and *select* the problems to be solved.

Hierarchy (of control)

The concept that control in a Human Activity System is exercised in a ranked or *hierarchical* sequence from the top-level downwards.

Hierarchy (of systems)

The principle that any system will comprise a series of sub-systems, each displaying system characteristics and the properties of whole bodies.

Holistic

Displaying the characteristics of a whole body or system.

Human Activity System

A notional system (ie not existing in any tangible form) where human beings are undertaking activities that achieve some purpose.

Information categories

The information needed, or produced, by a variety of system activities, assembled into generic groups (eg financial information, personnel information etc).

Input

That which enters a system in tangible or abstract form, and is changed or transformed by the system into an *output*.

Instruments

A pre-packaged set of tools, techniques and application principles which are utilised within the *FAOR* framework to address a situation from a desired viewpoint.

Interaction

The means by which the components of a system relate to other system components, ie by the transfer of *inputs* to, and *outputs* from, sub-systems.

Interview analysis

A specific process whereby points of commonality from interview findings are grouped together, either manually or by making use of a computer database.

Issues

Fundamental concerns about a situation illustrated as part of a rich picture; also taken to include concerns that are frequently expressed during an investigation.

Issue-based root definition

A root definition of a system that reflects a view that could be considered contentious (eg a prison as a *punishment* system).

Loose brief

A directive agreed for a study that does not stipulate a particular problem to be addressed, but is concerned with making improvements generally eg *to undertake a departmental review, or a value-for-money study* etc.

Management Services

Formally defined in the British Standards document BS 3138 as *"Specialist groups or units established within organisations to assist and advise on improvements in executive management functions"*; taken to include the activities of Work Study, Organisation and Methods (O&M), Operational Research, ergonomics, and industrial engineering.

Measure of performance

A means of assessing the effectiveness of a system or procedure.

Method for Information Systems Enquiry (MINSE)

An approach that uses soft systems ideas to explore the information needs of an organisation.

Model of the Soft Systems Methodology

A diagrammatic representation of the seven stages of the Soft Systems Methodology.

Monitor and control element

Used to describe the component of a conceptual model that is concerned with setting and maintaining the standards of performance of a Human Activity System.

Natural systems

Configurations that occur naturally in the universe and can be regarded as displaying system characteristics and properties of emergence (eg animals, trees, plants etc).

Needs Analysis Instrument

Used within the *FAOR* framework to clarify the interests and preferences of people who will use a designed office system, and how they could affect its success.

Open systems

Systems that receive inputs from, and discharge outputs to, the *environment*.

Organisation

Within the context of this book, taken to mean *"any body of persons organised for some purpose"*.

Organisation and Methods (O&M)

A generic term covering the activities traditionally undertaken by Management Services practitioners to examine and improve the effectiveness of office workers, organisation structures, and clerical procedures.

Organisation structure

The formal arrangements of functions within an enterprise, normally displayed as an organisation structure chart.

Organisational analysis/review

The examination of the whole, or a part of, an enterprise to determine areas where improvements can be made to structures, staffing levels or procedures.

Output

That which leaves a system in tangible or abstract form, ie an *input* transformed by the system.

Owner

In a root definition of a Human Activity System, the person or persons to whom the system is answerable, and/or who could cause it to cease to exist.

Ownership hierarchy

The concept that a Human Activity System could have a number of *owners* defined at different levels, ie for the system as specified, or for the wider systems of which it is part.

Pragmademic gap

A phrase coined to describe the gap that could be said to exist between *pragmatists* who have had not had formal systems training, and those who have developed and applied systems ideas at a more *academic* level; used also to refer to the difference between idealised solutions generated during the systems thinking stage of the Soft Systems Methodology, and those that are attainable in practical terms.

Primary task root definition/system

A root definition of a system that reflects a neutral or obvious purpose of real-world enterprises (eg *manufacturing cars, selling newspapers* etc).

Problem owner

Formally defined as *"the person or persons taken by an investigator to be those likely to gain most from an achieved improvement in a problem situation"*. Regarded also as the person or persons who will be effective in making any changes that are proposed as a result of a study, or the client to whom the analyst is reporting.

Problem situation

A situation where there is felt to be problems, without a clear definition of what those problems are.

Procedure

The manner of proceeding in some action; taken in the *Procedure Audit* process to be a form of Human Activity System.

Procedure Audit

An approach that uses soft systems concepts to establish and monitor the effectiveness of a procedure.

Processes

The constantly changing factors in a situation that are illustrated in a rich picture.

Regulation

The process whereby system operation is controlled within preset parameters; in a Human Activity System, this control is exercised by a decision-making body or mechanism.

Resources

The means or assets, either tangible or abstract, that can be utilised by a system to achieve a desired transformation.

Rich pictures

A graphical summary of the main factors affecting a situation, reflecting factual information about *structures* and *processes,* subjective concerns or *issues,* and giving an impression of the overall *climate* or relationships.

Root definitions

A description of a Human Activity System that defines what it consists of and what it aims to achieve. A root definition is formulated by considering each CATWOE element, and forms the basis for a conceptual model.

Social/cultural systems

Groups of people (such as families, communities, nations, clubs and so on) viewed in systems terms.

Soft situation

A situation where behaviour is not precise or predictable, generally applied to areas of unstructured human activity. (See also 'unstructured situation'.)

Soft Systems Analysis

Formally defined as *"the use of systems ideas to analyse (ie examine) soft situations to identify where problems could exist"*. Also used as a generic term to cover all approaches that use systems ideas to examine organisations.

Soft Systems Methodology

A set of high-level guidelines for applying systems ideas to soft or unstructured situations, providing a general learning framework for problem-identification, normally prior to the application of problem solving techniques.

Strategic rules (for the Soft Systems Methodology)

Rules that assist an analyst to select the best approach for planning and applying the Soft Systems Methodology.

Structure

The slow-to-change factors of a situation that are included in a rich picture.

System

Formally defined as *"A group of things or parts working together or connected in some way as to form a whole"*; ie a configuration, or ordered arrangement, of components that interact and, as a whole, display emergent properties.

System characteristics

The traits that distinguish systems from other entities, as summarised in the formal systems model.

Systemic

Regarding things in systems terms, ie *as a system*.

Systematic

The act of being methodical, or approaching matters in an ordered sequence.

Tight brief

A directive for a study where the problem and type of solution has been clearly specified, eg *to advise on equipment purchases, reduce staffing levels, design an office layout,* and so on.

Transformation

The change that takes place within or because of the system (ie the conversion of *input* to *output*).

Unstructured situation

A situation where behaviour is not precise or predictable and the underlying relationships are not clear. (See also 'soft situation'.)

User requirement

A documented description of the functions that a computer system is required to support, providing the interface between potential users and system designers or suppliers. Essentially a non-technical document that does not pre-empt a technical solution or assume that particular equipment will be utilised.

Weltanschauung/worldview

The perspective of a situation that has been assumed when defining a Human Activity System, ie how it is regarded from a particular (explicit) viewpoint; sometimes described as *the assumptions made about the system..*

What/how division

The distinction between *what* is being accomplished by a system (eg the allocation of resources), and *how* this is achieved (eg deploying staff, issuing stores and so on).

Wider systems of interest

The system or systems of which a chosen Human Activity System is a sub-set, and which can affect it by exerting control, setting the rules for operation, and providing resources.

Work Study

A generic term covering the activities traditionally undertaken to examine and improve the effectiveness of manual workers, mainly concerned with time and motion exercises.

Index